Chemistry of Concentrated and Dried Milk

NIPA® GENX ELECTRONIC RESOURCES & SOLUTIONS P. LTD.
New Delhi-110 034

About the Author

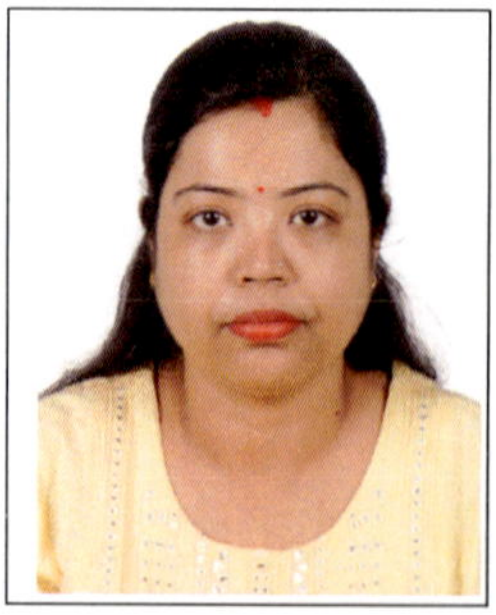

Dr. Chaitali Chakraborty has done B.Tech in Dairy Technology from West Bengal University of Animal and Fishery Sciences in the year 2002. She completed her Master's in Dairy Chemistry from West Bengal University of Animal and Fishery Sciences in the year 2004. Subsequently, she had done her PhD in Dairy Chemistry from West Bengal University of Animal and Fishery Sciences in the year 2008. She had worked in the department of Food Technology, Gurunanak Institute of Technology, MAKAUT, West Bengal as Assistant Professor. Currently she is working as Associate Professor in the department of Dairy Chemistry, Faculty of Dairy Technology, West Bengal University of Animal and Fishery Sciences, Nadia, West Bengal.

Chemistry of Concentrated and Dried Milk

Chaitali Chakraborty
Associate Professor
Department of Dairy Chemistry
Faculty of Dairy Technology
West Bengal University of Animal and Fishery Sciences
Mohanpur Campus, Nadia-741252
West Bengal, India

NIPA® GENX ELECTRONIC RESOURCES & SOLUTIONS P. LTD.
New Delhi-110 034

nipa®

NIPA® GENX ELECTRONIC RESOURCES & SOLUTIONS P. LTD.

101,103, Vikas Surya Plaza, CU Block
L.S.C.Market, Pitam Pura, New Delhi-110 034
Ph : +91 11 4386 0225, 9717133558, 9540816132
E-mail: newindiapublishingagency@gmail.com
Website: www.niparesources.com

Print ISBN: 978-93-5887-942-1

ebook ISBN: 978-93-5887-257-6

Composed and Designed by NIPA®.

Preface

The dairy processing sector is a rapidly growing industrial segment across the globe. To keep pace with this development, academic institutions are increasingly compelled to introduce disciplines related to Dairy and Food Science/Technology. In order to meet industrial demands, it is essential to regularly update curricula and research activities.

This book is primarily intended to support fresh graduates and postgraduates in the Dairy and Food Science/Technology fields. It aims to provide a solid foundation in theoretical concepts, thereby preparing students to effectively carry out laboratory experiments using various analytical techniques.

I am deeply grateful to all my well-wishers whose encouragement motivated me to undertake this endeavor. I extend my sincere thanks to my family members for their unwavering support and cooperation throughout the preparation of this book. I would also like to express my heartfelt appreciation to Ms. Ankita Pandit and Mr. Aniruddha Mukherjee for their kind support during the writing process.

Finally, I am thankful to the team at NIPA for their dedicated efforts in publishing this book in the interest of students. I warmly welcome suggestions from readers for any future improvements to this work.

Author

Contents

1

Chemistry of Concentrated Milk

1.1 Introduction

Concentrated milks, like evaporated milk and sweetened condensed milk, serve as practical substitutes for fresh milk in regions where access to it is limited or where transportation and storage are challenging. These products go through specific processing techniques to reduce their water content, thereby improving their shelf life and transportability.

Evaporated milk is created by removing water through evaporation and then sterilizing it to prevent microbial growth. This process results in a thicker consistency and a slightly caramelized flavor due to the heat treatment. On the other hand, sweetened condensed milk is made by adding a significant amount of sucrose to the milk and excluding oxygen, which inhibits microbial growth. This results in a thick, sweet product suitable for various culinary uses.

These concentrated milk products are usually packaged in small cans, making them convenient for storage and transportation. However, they are sometimes diluted before consumption with water to resemble the taste and consistency of fresh milk. In regions where even concentrated milks are not easily accessible, alternatives like whole milk powder or recombined milk are used. These alternatives involve dehydrating milk into powder form or reconstituting milk components to create a liquid similar to fresh milk.

Innovations in processing and packaging have led to the development of alternative uses for concentrated milks, expanding their versatility beyond traditional applications. These products continue to provide essential dairy nutrition in areas where fresh milk is scarce or hard to obtain, offering a shelf-stable solution to meet dietary needs.

1.2 History

Nicolas Appert introduced condensed milk in France in 1820, and Gail Borden Jr. replicated this achievement in the United States in 1853. By 1858, Borden's condensed milk, marketed as Eagle Brand, had earned a reputation for its purity, durability, and cost-effectiveness.

The inspiration for Borden's innovation stemmed from a tragic event: while returning from a trip to England in 1851, he lost several of his children, purportedly due to poor quality milk obtained from shipboard cows. Motivated to find a solution, Borden drew upon the vacuum pan technique utilized by Shakers to condense fruit juice. Adapting this concept, he developed a process for condensing milk without scorching or curdling it. This breakthrough transformed the dairy industry, enabling large-scale production of condensed milk while enhancing its shelf life, portability, and versatility. This development proved pivotal in food preservation, particularly in regions where fresh milk was scarce or challenging to transport. Borden's condensed milk quickly became a staple product, widely utilized in cooking, baking, and as a convenient source of dairy nutrition.

Condensed dairy products represent value-added milk items with an extended shelf life. Through processes like evaporation and occasional sugar addition, condensed milk becomes denser and more concentrated, thereby reducing its water content. This concentration effectively impedes microbial growth, prolonging the product's shelf life without necessitating refrigeration.

1.3 Definition

Condensed milks are derived by evaporating a portion of the water from whole milk, fully or partially skimmed milk, with or without the inclusion of sugar. These terms, "condensed milk" and "evaporated milk," delineate two distinct processed dairy products:

1. **Sweetened Condensed Milk:** This variant of condensed milk involves the concentration of full-cream milk coupled with the addition of sugar. It possesses a thick, creamy consistency and a heightened sweetness due to its elevated sugar content. Sweetened condensed milk finds widespread application in desserts, baking, and beverages.
2. **Evaporated Milk:** Also referred to as unsweetened condensed milk, this type of condensed milk entails the concentration of full-cream milk via evaporation to eliminate approximately 60% of its water content. Unlike sweetened condensed milk, it does not incorporate added sugar. With a thinner consistency than its sweetened counterpart, evaporated milk serves as a substitute for fresh milk in recipes demanding a richer flavor and creamier texture without the extra sweetness.

Both sweetened condensed milk and evaporated milk boast extended shelf lives compared to fresh milk due to their reduced water content, rendering them indispensable pantry items in many households. The milk solids concentration ratio stands at about 1:2.5 for full-cream products and 1:3 for sweetened condensed skim milk.

As per the Prevention of Food Adulteration (PFA) Rules of 1976, the specifications for various types of condensed milk are outlined as follows:

- **Unsweetened Condensed Milk (Evaporated Milk)**
    - This product is made by removing a portion of water from cow or buffalo milk, or a combination of both.
    - It may include minor additives like calcium chloride, citric acid, sodium citrate, and certain phosphoric acid salts (up to 0.3% of the total weight).
    - The milk should contain a minimum of 8.0% milk fat and 26.0% total milk solids.
- **Sweetened Condensed Milk**
    - This version is made by evaporating water from milk (cow or buffalo or a blend), followed by the addition of cane sugar.
    - Like unsweetened condensed milk, it may contain small quantities of additives such as calcium chloride and citric acid (not exceeding 0.3% of the finished product).
    - It must contain at least 9.0% milk fat, 31.0% total milk solids, and 40.0% cane sugar.
- **Unsweetened Condensed Skim Milk (Evaporated Skimmed Milk)**
    - This product is made by partially removing water from skim milk (cow or buffalo or a blend).
    - It may contain similar additives to unsweetened condensed milk (up to 0.3% of the product's weight).
    - It should have at least 20.0% total milk solids, with fat content not exceeding 0.5%.
- **Sweetened Condensed Skim Milk**
    - Prepared by removing water from skimmed milk and adding cane sugar.
    - Contains at least 26.0% total milk solids, 40.0% cane sugar, and fat content not exceeding 0.5%.

1.4 FSSAI Standards

Product	Milk fat	Milk solids	Milk protein in milk SNF
Sweetened condensed milk	Not lesser than 9% m/m	Not lesser than 31% m/m	Not lesser than 34% m/m
Sweetened condensed skimmed milk	Not lesser than 1% m/m	Not lesser than 26% m/m	Not less than 34% m/m
Sweetened condensed partly skimmed milk	Not lesser than 3% m/m and not more than 9% m/m	Not lesser than 28% m/m	Not lesser than 34% m/m
Sweetened condensed high fat milk	Not lesser than 16% m/m	Not lesser than 30% m/m	Not lesser than 34% m/m

1.5 Food and Nutritive Value of Condensed and Evaporated Milks

Condensed and evaporated milks are both rich in essential nutrients, including fat, fat-soluble vitamins (A, D, E, and K), proteins, minerals crucial for bone health, and lactose for energy. Sweetened condensed milk, owing to added sugar, is particularly high in energy from sucrose, while evaporated milk is gentle on digestion and suitable for infant feeding due to its soft curd formation.

During the sterilization process of evaporated milk, there is a loss of around 30-50% of vitamin B and approximately 60-100% of vitamin C. However, it's worth noting that milk itself isn't a significant source of vitamin C. Sterilization's impact on the biological value and digestibility of milk proteins is minimal.

1.6 Physico-chemical Properties of Concentrated Milk

1.6.1 Physical Properties

a) Evaporating water during the production of condensed milks increases their specific gravity or density, a parameter widely used to monitor their composition. Baume hydrometers are commonly utilized for this purpose.

b) Heat exposure during the manufacturing and storage of condensed milks can lead to color darkening and the development of a cooked flavor. This color alteration, known as browning or discoloration, occurs due to the interaction between amino compounds and sugars, primarily casein with lactose, termed Maillard-type browning. The resulting pigment is called melanoidin. The cooked flavor arises from the formation of sulphydryl compounds during heat treatment. These defects accumulate over time, with the rate and intensity of development being cumulative. Typically, the appearance of browning is accompanied by a cooked flavor, and vice versa. These reactions persist during storage.

The appearance of a brown colour in milk based products is undesired. While pasteurized milk generally doesn't undergo browning, certain condensed and dried milk products processed at high temperatures and stored at room temperature for prolonged periods are prone to this issue. Evaporated milk, in particular, is highly susceptible to such defects.

1.6.2 Chemical Properties (Maillard Browning Reaction)

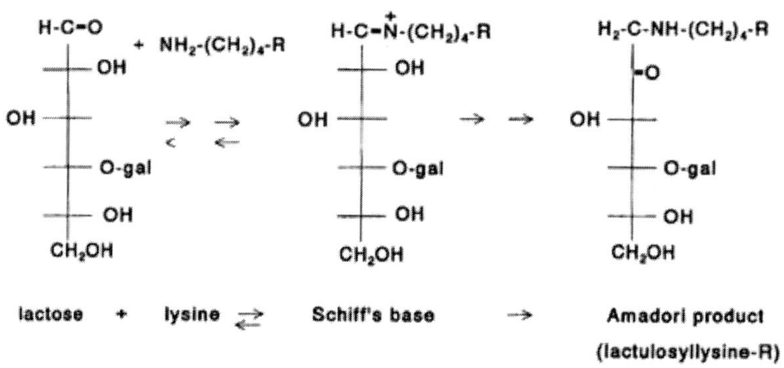

Fig. 1. Early Maillard reaction in milk, adapted from (Van Boekel 1998).

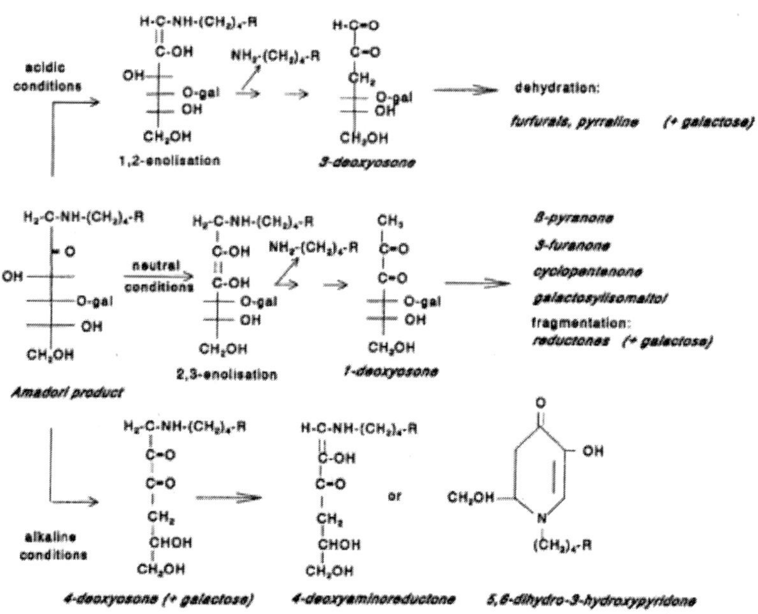

Fig. 2: Degrading Amadori compound under acidic, neutral and basic conditions, adapted from (Van Boekel 1998).

The major factors affecting browning-discoloration and cooked flavor production in fluid milk systems include:

(i) **Heat-treatment:** The intensity and duration of heating significantly influence the production of browning-discoloration and cooked flavors in milk. Higher temperatures and longer heating times can lead to more pronounced browning and development of cooked flavors, while gentler heat treatments result in less of these reactions.

(ii) **Total solids concentration:** The concentration of milk solids, particularly proteins and lactose, also plays a crucial role in browning and flavor development. Higher concentrations of milk solids, such as in condensed milk or milk powders, increase the potential for browning and cooked flavor formation when subjected to heat, compared to lower concentrations found in fresh milk.

(iii) **pH:** pH levels, often influenced by added stabilizers, impact browning and flavor development. Higher pH levels enhance these reactions, while lower pH levels inhibit them

(iv) **Storage temperature and time:** Higher storage temperatures and longer storage times increase the effects of browning and flavor development. Refrigerated storage tends to inhibit color development.

(v) Oxygen content:The oxygen content in the container's headspace influences browning and flavor development. Higher oxygen levels accelerate these reactions.

(vi) **Various added compounds:** Reducing sugars like lactose and glucose contribute more significantly to browning compared to sucrose. Additionally, non-amino browning or caramelization can occur due to heat-induced sugar decomposition.

(c) Viscosity: Viscosity control is crucial during the manufacture and storage of condensed milks. Several factors influence viscosity:

(i) **Concentration:** Higher concentrations of milk solids result in greater viscosity.

(ii) **Composition:** Increased percentages of casein, fat, and added sugar (in condensed milk) lead to higher viscosity.

(iii) **State of Protein and Fat:** Instability of milk proteins, particularly casein, can increase viscosity upon heating.

(iv) **pH and salt-balance:** Disturbed salt balance, often due to lower pH from acidity, increases viscosity.

(v) **Temperature of fore-warming/pre-heating:** Lower temperatures during pre-heating lead to greater viscosity.

Maintaining the right viscosity is crucial for successful marketing. Consumers associate viscosity with milk richness and nutritive value. Abnormal viscosity, either too low or too high, can lead to undesirable characteristics like sedimentation or gel formation. Thus, controlling viscosity ensures product quality and meets consumer expectations.

1.7 Basic technology of concentration

The manufacturing process of concentrated milk products relies on a fundamental requirement: the elimination of water. Modern technology offers three main methods to accomplish this:

1. **Evaporation:** Water is eliminated as a gas (vapor) through the process of evaporation.
2. **Reverse osmosis/ultrafiltration:** Water is extracted as a liquid through reverse osmosis or ultrafiltration processes.
3. **Freeze concentration:** Water is extracted as a solid (ice) using freeze concentration techniques.

1.7.1 Method of Manufacturing

Millk reception

↓

Filtration/ clarification

↓

Standardisation

↓

Preheating

↓

Sugar addition

↓

Condensation

↓

Allow to cool

↓

Crystallisation

↓

Package

↓

Storage

1.7.2 Basic component of an evaporator

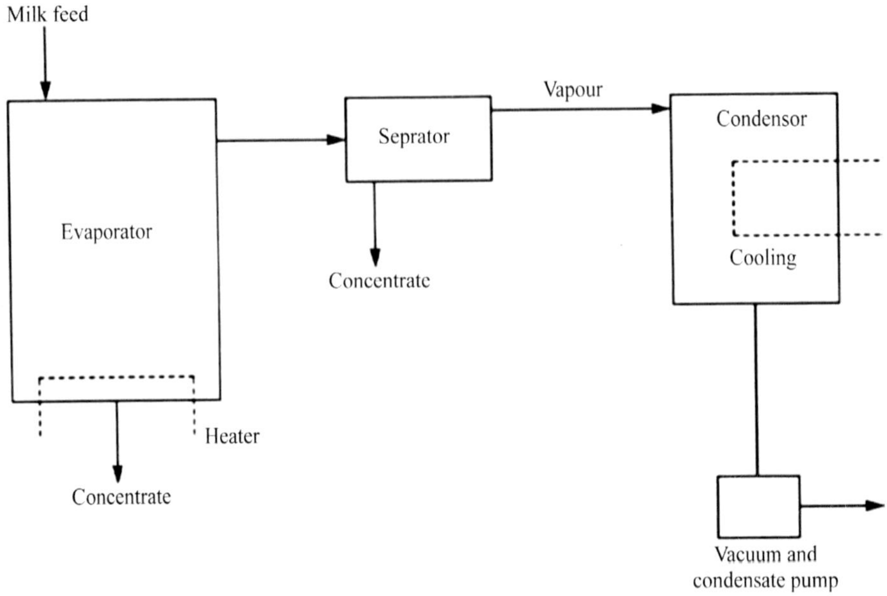

[Source: Varnam, 1994]

1.7.3 Changes Caused by Concentration

In addition to the rise in solute concentrations, the dehydration process of milk triggers various property changes, often corresponding to the concentration factor. These changes are affected by factors like heat treatment and homogenization. Key property alterations include:

1. Decrease in water activity as concentration increases.
2. Slight increase in calcium ion activity due to the transformation of saturated calcium phosphate into an undissolved state. As water content decreases, there's an increase in the association of ionic species and neutralization of protein ionic groups.
3. Modification in protein conformation due to changes in ionic strength, pH, and other salt equilibria. Higher concentration diminishes solvent quality, prompting protein molecule association and compact conformation. Casein micelles coalesce and enlarge, with a lesser increase observed if intense pre-heating has occurred, likely due to association with β-lactoglobulin and other serum proteins.

4. Elevation in osmotic pressure, freezing point depression, boiling point elevation, electrical conductivity, density, and refractive index, coupled with a decline in heat conductivity.
5. Increase in viscosity, transitioning the liquid into a non-Newtonian state and eventually solid-like.
6. Reduction in water diffusion coefficient from approximately 10^-9 m^2 s^-1 in milk to 10^-16 m^2 s^-1 in skim milk powder with minimal water content.

1.7.3.1 Effects on Viscosity

Stage of manufacture	Viscosity retardation
Fresh milk before pre-heating	15.24
Fresh milk after pre-heating	15.26
Evaporated milk before sterilisation	20.20
Evaporated milk just after sterilisation	150.00

1.7.3.2 Gel Formation During Storage

In certain instances of evaporated milks, viscosity gradually increases during storage, sometimes resulting in gel formation. This tendency is particularly notable in milks subjected to lighter heat treatments or with higher solids concentrations. Commercial observations, especially in tropical climates, support this phenomenon, particularly in high total solids (TS) evaporated milks.

Evaporated milks treated using the HTST process (135°C for 0.5 minutes) tend to thicken rapidly during storage. If the gel formation is interrupted before firming, the product may separate. Conversely, conventional sterilization methods (115.6°C for 20 minutes) result in slower gel formation that retains water better than the HTST process. As prolonged fluidity during storage is a significant marketable property, the more stringent heat treatment of the conventional process is commercially preferred over the less severe treatment of brief exposure to high sterilization heat.

The mechanisms behind age-related thinning or thickening are not fully elucidated. One possibility is that the HTST sterilization method does not allow sufficient time for attaining a new solubility equilibrium for calcium and magnesium salts, which are less soluble at higher temperatures. This could contribute to gel formation. Conversely, the conditions of the LTLT procedure (115.6°C for 20 minutes) may render more calcium and magnesium content insoluble, potentially encouraging protein particle swelling or continuation of the coagulation process, leading to irreversible gel formation.

1.7.3.3 Protein Separation on Storage

In certain cases, the proteins in evaporated milks have a propensity to settle during storage, a phenomenon more prevalent in evaporated skim milk compared to evaporated whole milk. Homogenization serves to delay protein separation in evaporated whole milk. Furthermore, it has been noted that the presence of fat in evaporated milk plays a role in maintaining a normal dispersion of milk constituents over extended storage periods. In this regard, fat serves to counterbalance the settling of proteins to the bottom of the container.

The protein found in evaporated milk is more easily digested compared to that in raw milk. It undergoes enzymatic hydrolysis more readily. Furthermore, when treated with rennet or pepsin, the casein in evaporated milk precipitates differently than in raw milk. While raw milk forms a solid curd, the casein in evaporated milk precipitates in a gelatinous form resembling a curd.

The unique behaviour of evaporated milk when treated with rennet may be partially attributed to the influence of heat, which can cause calcium to separate from the casein colloid complex. However, this explanation is not complete, as adding calcium ions does not restore the casein to its original curd-forming properties.

1.7.3.4 Changes in Colour

1. The colour of milk is significantly influenced by both the duration and temperature of heating.
2. Generally, during standard fore-warming periods, there is no notable change in milk colour. However, prolonged heating at higher temperatures (30°C) intensifies any effect.
3. Homogenization reduces the milk's colour by breaking down fat globules into smaller particles, preventing light from revealing the natural yellow hue of butterfat.
4. Milk gradually darkens over time and with increasing storage temperatures, although no colour change is observed at (50°C).
5. Prolonged heating can trigger Maillard browning, particularly if formaldehyde is added during sterilization, as it has a strong affinity for amino acids. The addition of formaldehyde maintains milk's white colour. Browning becomes more pronounced with higher storage temperatures, greater concentration during evaporation, and prolonged heating. Additionally, Maillard reactions may occur if the added sucrose contains invert sugar.

1.7.3.5 Occurrence of Cooked Flavour

Heat treatment during processing often imparts a cooked flavor to evaporated milk, a phenomenon also observed in fresh milk and other dairy products. It's commonly noted that the development of a cooked flavor coincides with color darkening. Compared to the LTLT process, the HTST process produces less cooked flavor. Additionally, HTST sterilization (135°C for 0.5 minutes) reduces the extent of sterilization, decreases viscosity, and can lead to fat separation, among other effects. Therefore, despite its limitations, the LTLT process remains important.

The reactions responsible for producing the cooked flavor are not fully understood. However, it's observed that heating skim milk, cream, and other dairy products can lead to the formation of -SH compounds from certain proteins, particularly when heated at high temperatures or for extended periods at lower temperatures. Nonetheless, -SH compounds are believed to be primarily responsible for the cooked flavor. A full cooked flavor can be achieved after five hours of heating at 70°C. It has been demonstrated that in the reaction causing the cooked flavor in milk due to heat, oxygen is consumed and carbon dioxide is produced. Oxidation conditions inhibit the formation of -SH groups.

1.7.3.6 Changes in Mineral Salt

In some instances, aged evaporated milk can develop granular deposits within the can. These deposits typically exhibit a whitish color, gritty texture, and insoluble nature, lacking crystalline characteristics. Upon analysis, these granular structures are found to contain tricalcium citrate and tricalcium phosphate, which have lower solubility in hot solutions compared to cold ones.

Evaporated milk manufacturers are well aware that an increase in milk solids concentration can lead to a higher amount of sediment formation. As the concentration rises, so do the levels of calcium, magnesium, citric acid, and phosphate in the milk, increasing the likelihood of precipitation. However, this can be managed by adding a casein stabilizer.

The formation of mineral salt precipitates as white, sand-like deposits is considered an age-related issue in evaporated milk. Storage temperature appears to be a significant factor in controlling this occurrence.

1.7.3.7 Effect of Storage Temperature

Storage Temperature	Time	Relative amount of Ca-citrate sediments
29.4° C	30 days	Large amount of fine as well as large particles
20.0° C	78- 110 days	Considerable amount of large particles
7.2° C	4 months	No sediments

1.7.3.8 Effect of storage, time & temperature on evaporated milk and condensed milk

The acidity level of condensed milk increases in direct correlation with both the duration and temperature of storage. Even after the initial heat treatment, the reaction that produces acid continues during storage, albeit at a slower rate depending on the storage temperature. For instance, sweetened condensed milk (SCM) with a sugar ratio of 64.5%, an initial acidity of 0.43%, and an original bacterial count of 4381/g displayed differing acidity increments when stored at various temperatures, as detailed in the table.

Temperature	Time	Acidity
10.0° C	640 days	0.84%
15.6° C	640 days	0.85%
22.2° C	640 days	1.0%

1.7.3.9 Fat separation in condensed and Evaporated Milk

During the phase of age thinning, preceding age thickening, there is a distinct tendency for undesirable fat separation. This highlights the crucial role of efficient homogenization. Effective homogenization, characterized by the absorption of casein onto fat globules, product viscosity, and minimal clustering of fat globules, is essential for controlling fat separation in recombined evaporated milk. Elevated ambient temperatures exacerbate this issue. Adding carrageenan provides optimal control over fat separation, although it may increase the risk of gelation. Ensuring the formation of a stable fat globule from anhydrous milk fat and solids-not-fat (SNF) is vital for recombined evaporated milk. Homogenization reduces fat globule size and increases viscosity, thus mitigating the tendency for fat separation.

1.7.3.10 Lactose Crystallization

Sweetened condensed milk typically contains about 38 to 45 grams of lactose per 100 grams of water. While lactose solubility at room temperature is around 20 grams per 100 grams of water, in sweetened condensed milk, this solubility reduces by about half due to the presence of sucrose. As a result, approximately 75% of the lactose tends to crystallize, meaning roughly 8 grams per 100 grams of sweetened condensed milk.

Due to the high viscosity of sweetened condensed milk, nucleation, the initial formation of crystals, occurs slowly, resulting in a few nuclei formed per unit volume of milk. This slow process leads to the development of large crystals unless special measures are taken. These crystals settle and contribute to a sandy mouthfeel. While individual crystals may not be large enough to be felt in the mouth, collectively, they create a sensation of roughness.

To address this issue, it's essential to ensure that the crystals formed are smaller than approximately 8 micrometers in length. While preventing crystallization entirely is not feasible, achieving a large number of smaller crystals is desirable. This can be achieved by using seed lactose. Adding 0.03% seed lactose represents 0.004 times the amount of lactose intended to crystallize.

The final crystal size in the product should not exceed 8 micrometers. Therefore, the seed lactose should contain enough seed crystals, ideally one per crystal to be formed, with a crystal size not exceeding approximately 1.25 micrometers. Such small crystals can be produced through intensive grinding of α-lactose hydrate.

1.8 Heat Stability of Milk and Condensed Milk

The coagulation of milk induced by heat remains a significant challenge in dairy chemistry, yet to be fully resolved. Heat coagulation is a result of a complex series of reactions, some of which complement each other while others antagonize. While much research has focused on the heat stability of un-concentrated milk, issues related to concentrated milks may hold greater commercial significance.

When milk is exposed to temperatures above its boiling point, it eventually coagulates, with higher temperatures leading to faster coagulation. The coagulation time is heavily influenced by pH, typically occurring at a low pH (<6.2). The heat coagulation time (HCT) at 140°C for bulk milk at its pH of maximum stability usually ranges from 20 to 30 minutes.

Casein, the primary milk protein, lacks secondary and tertiary structure and is not denaturable in the strict sense. However, it undergoes changes, primarily hydrolytic, when subjected to severe heat treatment. Casein micelles aggregate and undergo heat coagulation, unlike the heat denaturation experienced by globular proteins (whey proteins). Numerous heat-induced changes occur during prolonged heating at elevated temperatures, ultimately leading to coagulation. While some of these changes have been thoroughly investigated, others remain unstudied.

Some of the heat-induced changes in milk that can potentially lead to protein coagulation include:

1. Decreased value of pH
2. Precipitate of calcium phosphate formed.
3. Whey protein denaturation and its interaction with casein.
4. Maillard browning.
5. Modification of casein into dephosphorylation, hydrolysis of κ-casein, and general hydrolysis.
6. Changes in casein micelle structure:
   a. Zeta potential changes
   b. Hydration changes
   c. Association –Dissociation reactions.

1.8.1 Type of Milk

1. Type A milk refers to milk varieties characterized by a clear maximum (around pH 6.7) and minimum (around pH 6.9) in stability on the heat coagulation time (HCT) versus pH curve. Beyond a pH of 6.9, the HCT begins to increase again.
2. Type B milk describes milks where stability progressively increases with higher pH values, and there are no distinct maximum or minimum points on the HCT/pH curve. The specific shape of the HCT/pH curve is unique to each milk and typically remains consistent throughout lactation, likely influenced by dietary, environmental, and genetic factors.

pH stands out as a critical factor, with most bulk milks and individual cow milks exhibiting a maximum around pH 6.7 and a minimum around pH 6.9 on the HCT/pH curve.

The precise structure of the HCT/pH curve is effected by a whole lot offactors among which the important ones are:

1. Whey protein
2. Some heatdenaturableproteins
3. κ-casein
4. Colloidal calcium phosphate complex
5. Detergents
6. Assay conditions (temperature & agitation)
7. Urea

8. Preheating/Fore warming
9. Aldehydes
10. Concentration

1. Below the pH maximum, the coagulation of casein micelles is primarily governed by the calcium-induced complexation of casein. As pH increases, the concentration of calcium ions decreases, leading to increased heat stability of casein micelles. However, once protein polymerization, dephosphorylation, and hydrolysis of κ-casein occur, reducing casein stability, the lower concentration of calcium ions becomes sufficient to induce coagulation. The presence of the heat stability minimum (around pH 6.9) is attributed to the formation of β-lg and κ-casein complexes, sensitive to calcium phosphate precipitation, and increased heat-induced calcium phosphate deposition.
2. Acid produced during heating may be the main cause of coagulation at the heat stability maximum, and periodic neutralization significantly delays coagulation.
3. The transition from the coated (by denatured whey protein) to the protein-depleted (depleted of κ-casein) form, occurring as pH increases above the maximum heat stability pH, results in a fluctuation in micellar heat stability. Initially, there's an increase (the maximum), followed by a decrease (the minimum), and then another increase in heat stability over the pH range of 6.4-7.0.
4. The occurrence of the minimum in the HCT/pH profile is attributed to the heat-induced dissociation of micellar κ-casein, which happens when milk is heated beyond 90°C at pH values above 6.9. Micelles depleted of κ-casein are sensitive to calcium phosphate precipitation, leading to the heat stability minimum. The high heat stability at pH values above the minimum is likely due to the availability of calcium ions (resulting from heat-induced precipitation of calcium phosphate) and the high net negative charge on the protein. Therefore, inducing coagulation at pH values above ~7.2 requires longer heating times to cause sufficient κ-casein dissociation, hydrolysis, pH reduction, or other heat-induced changes to decrease the micellar negative charge enough for coagulation.

Microstructural changes observed in electron micrographs of skim milk heated at 130 or 140°C include:

1) A reduction in the range of micelle size and an increase in the average micellar diameter.

2) Below pH 6.7, heating induced changes in the micelle surface occurs from smooth to a ragged one with many appendages, attributed to ß-lg / κ-casein complexes on the micellar surfaces. Additionally, more ß-lg precipitates onto the surface layer during heating.
3) Above pH 7.0, ß-Lg / κ-casein complexes do not form upon heating milk.
4) At the pH of maximum heat stability (pH 6.7), the formed coagulum displays aggregated casein micelles in short chains within a relatively open network, with many individual micelles still present.
5) At the pH of minimum heat stability (pH 6.9), milk coagulum exhibits a closed network of long chains of fused micelles. Casein micelles join in chains and clusters, gradually building into a gel network during the last few minutes before visible coagulation. Post-coagulation, the gel network becomes denser and comprises long chains of fused micelles, indicating an abrupt change at coagulation.

1.8.2 Additives Used

Urea and lactose don't notably enhance the stability of concentrated milk (25% TS) individually, but they do exhibit synergistic effects with aldehydes. Simple aldehydes, when added, significantly boost the heat stability of milk across a broad pH range. Certain sugars, which function as aldehydes upon heating, also contribute to stabilizing concentrates at high temperatures (120°C).

Alkalis and salts like sodium hydroxide (0.05 to 0.10%) and tri-sodium phosphate (0.075%), added before preheating, enhance the heat stability of skim milk concentrates with 18.5% to 22.5% total solids. Conversely, calcium chloride has an opposite effect.

To ensure heat stability in concentrated milk, stabilizing techniques must be applied prior to homogenization. The primary impact of stabilizers arises from their influence on the milk's pH.

In buffalo milk concentrates, the greatest increase in heat stability occurs in preheated (120°C, without holding) concentrates (35% TS) with the addition of disodium hydrogen phosphate at a 0.3% level. Tri-sodium citrate, added at the same level, also enhances stability, albeit to a lesser extent. Additionally, added casein contributes somewhat to the heat coagulation time of buffalo milk concentrate.

1.9 Age Thickening and Gelation in Condensed Milk

As evaporated milk is stored, its viscosity might initially decrease slightly. This decrease can be attributed to changes in the shape of casein micelle aggregates formed during sterilization, transitioning from irregular to spherical shapes. Consequently, the effective volume fraction decreases. However, with time, the viscosity of the milk increases significantly. This increase in viscosity becomes highly dependent on the shear rate, with the milk displaying a yield stress. Eventually, a gel forms, which firms rapidly.

1.9.1 Mechanism of Age-gelation

The precise mechanism behind age thickening or gelation in evaporated milk remains uncertain. It's generally not attributed to proteolytic enzymes or Maillard reactions, although these processes may occur concurrently with gelation. Heat coagulation doesn't seem to play a significant role, nor does pH have a pronounced effect. Gelation tends to accelerate with decreased calcium content, and electron microscopy reveals thread-like protrusions on casein micelles, eventually forming a network. While changes in micellar calcium phosphate likely contribute, a definitive explanation is lacking.

Age thickening and gelation occur more rapidly in UHT-evaporated milk, possibly due to proteolysis by psychrotrophs, even in the absence of such enzymes. Stronger sterilization post-evaporation delays gelation, while higher milk concentration and storage temperature hasten it. Sodium polyphosphate delays gelation significantly, while citrate or orthophosphate may accelerate it by binding calcium. Polyphosphates may hydrolyze into orthophosphate, particularly during heating, so their addition doesn't necessarily prevent gelation in bottle-sterilized evaporated milk.

Conventionally processed evaporated milk typically only gels after prolonged storage at high temperatures, often accompanied by extensive Maillard reactions. Rapid gelation may occur if refrigerated before sterilization, particularly at 40°C for a few days.

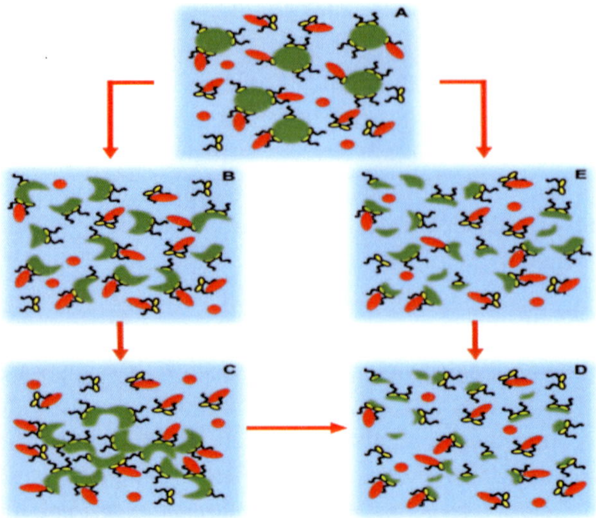

Fig. 1: Age gelation of milk by indigenous proteases (plasmin)
[Source:Harwalkar, 1992]

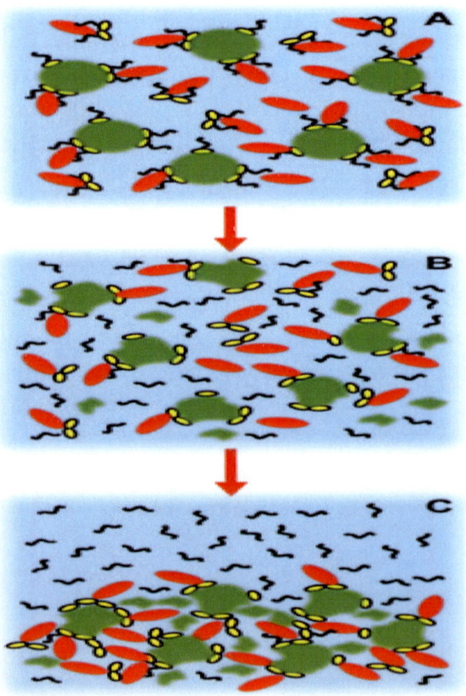

Fig. 2: Age gelation of milk by exogenous proteases (bacterial protease)
[Source: Harwalkar, 1992]

1.9.2 Process of Age Gelation in UHT Milk

(A) In UHT milk, both denatured and undenatured whey proteins may be present, with denatured whey proteins found either in the serum phase or attached to κ-casein on the casein micelles.

(B) Plasmin, a proteolytic enzyme, gradually hydrolyzes casein micelles, β-casein, and alpha S1-casein/alpha S2-casein, resulting in the generation of peptide fragments.

(C) These peptide fragments interact through hydrophobic bonding and calcium bridges, leading to the formation of a gel network.

(D) Further breakdown of the casein by plasmin causes the gel to disintegrate, resulting in the clarification of the skim milk.

(E) Plasmin-mediated rapid hydrolysis of the casein micelles produces small fragments that do not contribute to gel formation but instead contribute to the clarified skim milk.

This process illustrates how the action of plasmin on casein in UHT milk leads to the formation and subsequent breakdown of gel structures, ultimately affecting the texture and stability of the milk product.

1.10 Defects in Sweetened Condensed Milks

1. Sandiness

In sweetened condensed milk typically arises from the size of lactose crystals present. The texture can be smoother if the crystals are smaller, ideally averaging ≤10μ in length. Excessive sucrose content, particularly above 64.5%, can lead to a saturation point resembling a saturated solution of sucrose. Upon exposure to low temperatures during storage or transit, excess sucrose may crystallize, resulting in a coarse, sandy texture. Additionally, incomplete solution during processing can contribute to sandiness.

When sweetened condensed milk is allowed to cool excessively before packaging and then reheated with agitation, sandiness may develop. Vigorous agitation during reheating can cause smaller lactose crystals to dissolve again, increasing super-saturation and promoting the growth of remaining crystals. This results in larger aggregates, giving the milk an undesirable sandy texture.

2. Sugar Sediment or Sugar Down

Sugar sedimentation in sweetened condensed milk often occurs, with microscopic examination revealing lactose crystals as the primary component. This sedimentation is primarily caused by the disparity in specific gravity between crystallized lactose and the rest of the condensed milk. The specific gravity of lactose hydrate at 68°C is notably higher than that of sweetened condensed milk. Notably, the size of lactose crystals plays a significant role, with larger crystals leading to more pronounced sedimentation.

Viscosity also influences sugar sedimentation. Higher viscosity results in increased resistance to gravity, reducing the likelihood of lactose crystals settling at the bottom. In cases where sedimentation is problematic, the finished product may be too thin to suspend crystals adequately. Increasing the concentration of the condensed milk can mitigate sedimentation, provided that crystal size is consistently controlled.

3. Age Thickening

Age thickening in sweetened condensed milk can stem from both bacterial activity and physicochemical reactions. Bacterial thickening arises from severe bacterial contamination, often due to inadequate sanitary oversight during manufacturing. This issue can be mitigated through stringent sanitation practices and precise control of fore-warming temperatures. Adjusting the sucrose-to-water ratio to approximately 64.5% can help control bacterial thickening in the presence of causative organisms.

Physicochemical age thickening, on the other hand, involves hydration of proteins, particularly casein, which absorbs water and contributes to increased viscosity. The influence of seasonal variations on age thickening is significant, highlighting the importance of proper fore-warming temperature adjustment to prevent this defect. While the exact mechanisms underlying physicochemical age thickening remain unclear, addressing these factors can help minimize the occurrence of this issue in sweetened condensed milk production.

4. Adjustment of Sucrose Content

Modifying the ratio of sucrose to water can influence the tendency for age thickening. Increasing this ratio reduces the likelihood of thickening over time. By adding more sucrose to fresh milk during the production of sweetened condensed milk, manufacturers can decrease the viscosity of the concentrated milk, thus delaying or preventing age-related thickening. This ratio may need seasonal adjustments to maintain product stability.

5. Controlling Pan Temperature

During the milk condensation process, the temperature of the pan plays a critical role. Higher temperatures can destabilize the product and increase the risk of age-related thickening. Lowering the pan temperature toward the end of condensation can effectively reduce viscosity and mitigate the likelihood of thickening.

6. Managing Solids-Not-Fat (SNF) Content

The concentration of solids-not-fat (SNF), particularly in relation to water content, can impact age thickening tendencies. Higher SNF levels, if not properly managed, can increase the likelihood of thickening. Adjusting the SNF content before processing can help minimize this risk.

7. Maintaining Fat-to-SNF Ratio

Maintaining an appropriate ratio, such as 8.5% fat to 19.5% SNF according to the U.S. Standard, can help mitigate factors contributing to protein colloidal swelling. This ratio is crucial as fat acts as a softener and dilutant in the protein suspension, resulting in a smoother texture compared to products made from skim milk. Sweetened condensed skim milk, lacking fat, is more susceptible to age-related thickening.

1.11 Defects in Evaporated Milks

Evaporated milk, in order to be marketable, must be free from symptoms of biological fermentation. It must be sterile. In addition, it must have certain physically important properties such as full of body that symbolizes richness and prevents objectionable fat separation, smooth and homogeneous texture, a pleasing natural milk flavour and a light creamy colour.

1. Curdiness Due to Heat Coagulation

For the dependable prevention of curdiness and other physical defects that dissipate the marketable properties of the finished product, it is necessary to efficiently control the viscosity of the evaporated milk. The causes and prevention of curdiness due to heat coagulation are briefly summarized as follows:

A. Curdy evaporated milk is due to too low heat stability in the presence of heat treatment used for sterilization.

B. The standard commercial sterilizing process of holding at 115.6°C for 20 min or other temperature time ratios with equal lethal effect represents the minimum temperature time ratio that will insure a sterile finished product.

C. When the cooking time (minutes held at full sterilizing temperature) approaches the heat stability time, there is danger of permanent curdiness in the finished product.

D. If the heat stability of the evaporated milk is too low to ensure freedom from curdiness, it may be increased and curdiness prevented by use of higher forewarming temperature or by the proper use of casein stabilizer.

E. Too old or otherwise inferior raw milk as well as holding the evaporated milk too long or at too high a temperature before sterilization are not the frequent causes of curdy evaporated milk.

F. Higher homogenizing pressure such as pressure between 211 & 281 kg/cm^2 or higher lowers the heat stability and increases the tendency to curdiness. Raw milk of inferior quality or condensed milk held improperly before sterilizing or homogenizing pressure in excess of 175 kg/cm^2 increases the tendency to a curdy finished product.

2. Fat Separation in Storage

Adoption of homogenization as an integral function of manufacturing of evaporated milk eliminates the defect of fat separation as a serious problem. Yet even in the homogenized product, the distribution of fat globules is not permanently uniform, on progressive storage of homogenized product, it becomes higher and richer on top layer.

If objectionable intensity of fat separation is to be avoided, it is necessary in addition to efficient homogenization (the fat globules uniformly to a diameter of 2 μ or less) to

(i) Aim not for maximum heat stability but for heat stability sufficiently moderate to give evaporated milk a full body and a viscosity adequate to definitely retard fat separation in storage.

(ii) To accomplish this, the processor aims at a concentrate that possesses heat stability a few minutes longer than about twice that of the sterilizing period.

(iii) Varying the sterilizing process, yields the optimum viscosity for maximum fat emulsion stability.

3. Mineral Deposit

Under certain conditions, there is a tendency for a whitish gritty deposit to form near the bottom of the container. This deposit consists chiefly of tri calcium citrate. Its causes & prevention are as follows:

(i) Citric acid is believed to be present in milk combined with Ca and Mg. The citric acid content of milk has been found to be greatest when the cows are on green pasture.

(ii) Higher the concentration of evaporated milk, the greater the tendency for appearance of a mineral deposit during storage.

(iii) The temperature of storage is a controlling factor. The higher the temperature, the larger the amount of deposit. At 7.2°C there is no mineral deposition.

4. Browning of Evaporated Milk

The reasons for browning are discussed somewhere else. However, the causes and prevention are summarized as follows:

(i) The darkening of colour is due to reaction in the presence of sterilizing heat, between the sugar of milk and certain amino acid. This reaction causes discoloration of the proteins.

(ii) The Browning effect is much less intense in the case of high temperature short time heat treatment then LTLT.

(iii) The colour of evaporated milk darkens with age. The rate of darkening increases with storage temperature. At 5°C or below there is no change of colour during storage.

5. Bloats of Evaporated Milk

The bloats are caused due to

(i) Freezing,

(ii) Change in altitude and

(iii) Chemical actions.

Bloats resulting from these causes are frequently occurring. When freezing, the content expands sufficiently to cause the cans to buldge at their ends, giving the impression of bloats. When subsequently transferred to temperatures at which the milk melts, the buldging ends almost invariably flip back to normal.

Change to high altitude may cause tins to buldge. This will result especially when cans are sealed at low altitude and are then transferred to high altitude regions. The danger in such cases is intensified, if the evaporated milk happens to be cold at times of filling and atmosphere at high altitude is warmer causing the air and milk in can to expand. The pressure difference between inside and outside of tins thus causes the can ends to buldge out.

Chemical action of milk acid upon the base metal of the container affects the appearance of bloats.

2

Chemistry of Dried Milk

2.1 Introduction

The continuous production of dry milk has become consistently significant within the dairy industry and is expected to continue growing due to its advantageous features. These include improved shelf life, reduced storage requirements, and lower transportation costs, all of which contribute to favourable economic outcomes.

For example, non-fat dry milk is also known as skim milk powder, dried skim milk, non-fat dry milk solids, or dehydrated skim milk. Similarly, products like dried whole milk, dried cream, and dried buttermilk are often regarded as dried or powdered whole milk, cream, and buttermilk, respectively, based on their specific compositions.

2.2 Types of Dried Powder

1. Nonfat dry milk
2. Dry whole milk
3. Dry buttermilk
4. Dry whey
5. Dry ice cream mix
6. Dry creams (sweet or sour)
7. Dry skim milk vegetable fat products
8. Malted milk Powder
9. Cheese powders (Cheddar, Blue, etc.)
10. Sweetened chocolate flavoured non fat dry milk
11. Coffee creaming products
12. Sodium or Calcium Caseinates
13. Dry high acid milks
14. Partly skim milk powder
15. Infant milk food

16. Lactose powder
17. De-lactose milk powder

2.3 FSSAI Standards

Parameter	Whole milk powder (WMP)	Partly skimmed milk powder	Skimmed milk powder (SMP)	Cream powder
Moisture, maximum, % (m/m)	5.0	5.0	5.0	5.0
Milk fat, %, (m/m)	Minimum 26.0 and lesser than 42.0	Greater than 1.5 and less than 26.0	1.5 (maximum)	42 (minimum)
Milk protein present in milk solid- not-fat, minimum, %, (m/m)	34.0	34.0	34.0	34.0
Titratable acidity, maximum (ml 0.1 NaOH for 10 g solid-not-fat)	18.0	18.0	18.0	–
Insoluble index, maximum, ml	2.0	2.0	2.0	–
Total ash, maximum, % (m/m), on moisture and fat free basis	9.3	9.3	9.3	–

2.4 Flow-diagram for Manufacturing of Dried Milk

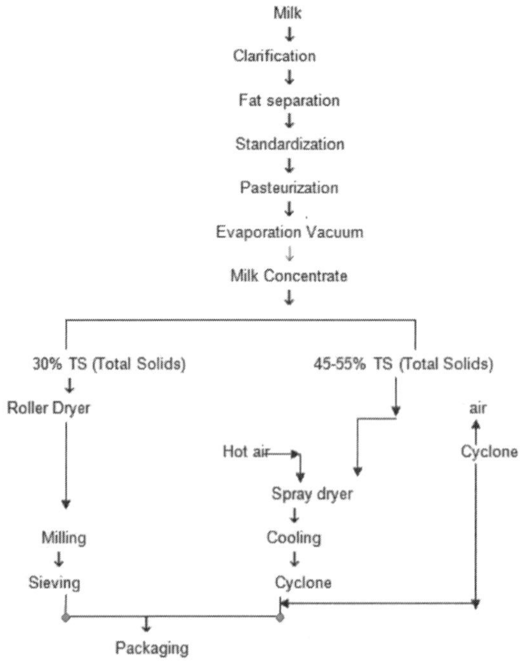

(*Source*: TNAU Agritech Portal; Post Harvest Technology)

2.5 Basic Technology of Drying

There are two primary methods for removing water in a "free" state:

1. Evaporation: This method is utilized in roller drying. Here, the saturated vapor pressure of the surface releasing moisture equals atmospheric pressure (P"vo = P).
2. Vaporization: This process is employed in spray drying. In this method, the saturated vapor pressure of the surface releasing moisture is less than atmospheric pressure (P"vo< P).

Spray drying typically involves combining atomized milk with hot air in a drying chamber. The droplet diameter is usually less than 300 micrometres, preferably between 10 and 100 micrometres. The hot air, with an inlet temperature ranging from 150°C to 220°C, transfers heat to the droplets, causing moisture to evaporate. The vaporization temperature, at commonly used inlet temperatures, falls between 40°C and 50°C. It's crucial to ensure that the temperature of the dry particles never exceeds that of the cooled outlet air. Finally, the dry particles are separated from the air, either before or after exiting the drying chamber, and cooled before being packaged.

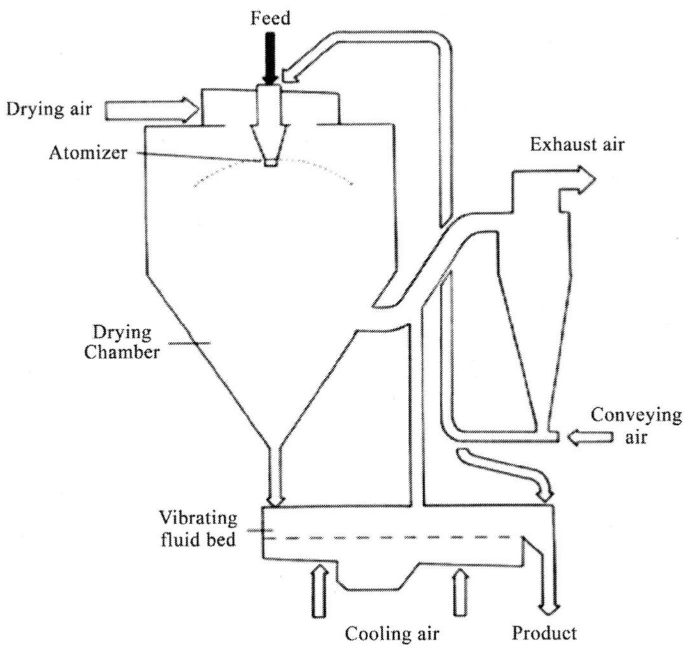

Fig. 3: Two stage spray drier
[*Source*: Varnam, 1994]

2.6 Changes Occur During Processing

1. Fore-warming

Standardized milk or skim milk undergoes preheating to fulfill specific product criteria regarding whey protein denaturation. For instance, high-heat milk powder must contain 1.5 mg of undenatured whey proteins per gram, while low-heat milk powder should contain 26.0 mg of undenatured whey proteins per gram.

Low-heat milk powder is derived from milk subjected to fore-warming at precise temperature and duration, such as 72° to 74°C for 15 seconds, ensuring that less than 10% of whey protein is denatured.

2. Vacuum Evaporation

Fore-warmedskim milk is concentrated to achieve a total milk solids content of 40-60%through vacuum evaporation. The concentration level is determined based on dryer design specifications and is utilized to regulate the particle size, density, dispersibility, and solubility of the resulting powder.

2.7 Instantization

Non-fat dry milk (NFDM) commonly used in homes and institutions often undergoes instantization to enhance the powder's dispersibility. This process involves partially rehydrating spray-dried milk powder with steam until it reaches a moisture content of 11 to 16%, converting lactose into its monohydrate form. This transformation makes the powder sticky, facilitating particle agglomeration. The agglomerated milk powder particles are then dried again using a fluidized bed dryer. Some specially designed spray dryers allow for simultaneous drying and agglomeration of the particles to produce instantized milk powder. These agglomerated particles are subsequently removed from the spray dryer, and the drying process is completed on a belt dryer.

Instantized milk powder typically has a lower bulk density but better water dispersibility compared to conventionally spray-dried powder. However, the additional heat exposure during agglomeration and re-drying can lead to further Maillard reactions, whey protein denaturation, and related chemical and physicochemical reactions, which may degrade product quality.

Instantization is the process through which dried milk and milk products are rendered instant-soluble.

2.8 Properties of Milk Powder

1. **Flowability:** Flowability is the capacity of a powdered substance to flow, determined by factors like particle shape, size, intermolecular forces, and porosity.
2. **Wettability:** Wettability measures a powder's ability to be wetted by water at a specific temperature, particularly crucial in instant powder production. It depends on whether particle surfaces are water-repellent or absorbent, impacting how quickly they form a film when in contact with water. Fine powders from spray processes can be challenging to wet and may form lumps, with particles under ~50μ swelling upon initial water contact, hindering water penetration. An ideal particle size is typically around 100-150μ.
3. **Sinkability:**Sinkability refers to the tendency of wetted particles to sink in water due to milk component dissolution and dispersion. Particle density, influenced by composition and occluded air, affects sinkability. During reconstitution, particles initially sink, but if their density becomes equal to or lower than the liquid, they may rise again. To prevent this, particle density should remain high, minimizing occluded air content.
4. **Dispersibility:** Dispersibility is the ability of a powder to disperse in water upon gentle stirring, crucial for instant powders. Optimal dispersibility requires good wetting properties and optimal agglomeration to avoid the presence of fine particles. Dispersibility is often assessed using various methods, with the IDF considering powders instant if dispersibility is at least 85% for whole milk and 90% for skim milk. Modern drying technologies can achieve dispersibility levels of 95%.
5. **Solubility:** Solubility refers to a powder's ability to dissolve in water, crucial for achieving fully reconstituted milk quickly and with minimal effort. Capillary penetration of water into the powder must be prevented or delayed to allow for dispersion before dissolution. Solubility depends on factors like agglomerate size, non-agglomerated particle size and amount, interstitial air, and specific surface area. While not all powder components are soluble upon reconstitution, modern drying techniques minimize insolubility, approaching nearly 100% solubility. Heat-induced protein denaturation during manufacturing can affect solubility, making it an essential commercial property.

3

Chemistry of Malted Food

3.1 Malted Milk

Malted milk, originating from the ideas of British pharmacist James Horlick in the 19th century, was initially marketed under the name "Diastoid" before being trademarked as "malted milk" in 1887. Although initially targeted for infants and invalids, its appeal expanded due to its nutritional value, convenience, digestibility, and palatability. Recognized globally for its lightweight, non-perishable nature, and high-calorie content, malted milk is created by evaporating a mixture of malted barley, wheat flour, and whole milk until it forms a powder.

There aremainly two types of malt powder: diastatic and non-diastatic. Diastatic malt containing enzymes breaking down starch into sugar, aiding in dough rise and crust formation, particularly in bread making. Non-diastatic malt lacks active enzymes and is primarily utilized for flavor enhancement, commonly in beverages. It may also include sugar, coloring agents, and other additives to improve taste and appearance.

3.2 Malting

Malting is a crucial process employed in treating cereal grains, such as barley, by inducing germination through soaking in water followed by halting the process through drying or heating with hot air. This technique encompasses two main stages: sprouting and kiln-drying. The term "malt" encompasses several products of this process:

1. **Malted grains:** These are grains, like barley, that have undergone the malting process.
2. **Malt sugars:** These are sugars, predominantly maltose-rich, derived from malted grains, often utilized in baking, such as baker's malt.
3. **Malted milk products:** These include beverages like malted milkshakes, made using malted milk.

Malting grains activates enzymes essential for converting starches in the grain into fermentable sugars, including monosaccharides such as glucose and fructose, as well as disaccharides like sucrose. Furthermore, it fosters the development of enzymes like proteases, which break down proteins in the grain into forms usable by yeast during fermentation. Barley is a flavoured choice for malting due to its high enzyme content, known as diastatic power, necessary for these enzymatic processes.

3.3 Drying Malted Milk

The manufacturing of malted milk foods typically involves the tray-drying method under vacuum, which is a batch process and can be costly. In contrast, the spray drying method offers a more economical solution due to its continuous and automated operation at a larger scale. Therefore, transitioning to the spray drying method can potentially produce malted milk food with enhanced physical attributes at a reduced cost compared to the traditional tray-drying technique.

3.4 Malted Foods

1. Malted milk foods
2. Malt based foods

Malted milk food is a combination of whole milk, partly skimmed milk, or milk powder blended with wort derived from a mash of ground barley malt, along with other malted cereal grains and wheat or other cereal flours. This mixture may also contain flavorings, spices, emulsifiers, eggs, protein isolates, edible common salt, sodium or potassium bicarbonate, minerals, and vitamins. It is processed into powder, granules, or flakes using methods like roller drying, spray drying, or vacuum drying. While cocoa powder may be incorporated, the product must be free from contaminants. It should not include added starch (except for the starch naturally found in cocoa powder) or non-milk fats. Additionally, it must not contain preservatives or artificial colors, although products with cocoa powder may contain added sugars. Malted milk food must also meet the following standards:

3.5 FSSAI Standards

Parameters	**Malted milk food without cocoa powder**	**Malted milk food with cocoa powder**
Moisture	Not more than 5% by weight	Not more than 5% by weight
Total protein (on dry matter basis)	Not less than 12.5% by weight	Not less than 11.25% by weight
Total fat (on dry matter basis)	Not less than 7.5% by weight	Not less than 6% by weight
Total ash (on dry matter basis)	Not more than 5% by weight	Not more than 5% by weight
Acid insoluble ash (on dry basis in dil. HCl)	Not more than 0.1% by weight	Not more than 0.1% by weight
Solubility	Not more than 85% by weight	Not more than 80% by weight
Cocoa powder (on dry basis)	_	Not less than 5% by weight
Test for starch	Negative	_
Bacterial count	Not more than 50,000 per gram	Not more than 50,000 per gram
Coliform count	Not more than 10 per gram	Not more than 10 per gram
Yeast & mould count	-	Absent in 0.1 gram
Salmonella and *Shigella*	-	Absent in 0.1 gram
E. coli	-	Absent in 0.1 gram
Vibrio cholera	-	Absent in 0.1 gram
Staphylococcus aureus	-	Absent in 0.1 gram

Malt-based foods, or malt food, are products made by combining malt (such as wort, flour, or malt extract) obtained from the controlled germination of seeds, including cereals and legumes, with other cereal and legume flours. The production process includes steeping, germination, and kiln drying. Additional ingredients, such as whole milk, milk powder, flavorings, spices, emulsifiers, eggs, egg powder, protein isolates, protein hydrolysates, edible common salt, liquid glucose, sodium or potassium bicarbonate, minerals, amino acids, and vitamins, may be added. Sugar and/or cocoa powder might also be included.

These products are typically processed into powder, granules, or flakes through drying or dry mixing of the components. The cereals, legumes, and their derivatives used in making malt must be of high quality, free from infestations, and devoid of contaminants like insect fragments, rat droppings, or mold. The final product must meet the following standards:

3.6 FSSAI Standard

Moisture	**Not more than 5% by weight**
Total protein (on dry matter basis)	Not less than 7% by weight
Total ash (on dry matter basis)	Not more than 5% by weight
Acid insoluble ash (in dil. HCl)	Not more than 0.1% by weight
Total plate count	Not more than 50,000 per gram
Coliform count	Not more than 10 per gram
Yeast & mould count	Not more than 100 per gram
E. coli	Absent in 10 gram
Salmonella and Shigella	Absent in 25 gram
Alcoholic acidity	Not more than 0.3% by weight
Vibrio cholera	Absent in 0.1 gram
Staphylococcus aureus	Absent in 0.1 gram

4

Chemistry of Human Milk and Infant Food

4.1 Introduction

Human milk widely acknowledged as the optimal source of nutrition for infants, providing a rich array of components crucial for their growth, repair and development. However, the composition of human milk can vary significantly within and between feeds, throughout the day, over the course of lactation, and among different mothers and also populations. While this variability is beneficial for infant health and survival, it presents challenges for the management of nutrition of high-risk infants who require precise oversight.

The composition of human milk can be altered by factors such as the handling of expressed milk, including how it is stored and pasteurized. Thus, proper management of expressed milk is essential. The recommended practice for infant nutrition is exclusive breastfeeding for the first six months, followed by continued breastfeeding for at least one to two years. Human milk is naturally suited to meet the nutritional needs of infants, providing both nourishment and bioactive compounds that support healthy growth and immunity.

Human milk contains a variety of bioactive components, such as immune cells, anti-infective and anti-inflammatory agents, growth factors, and prebiotics. In contrast to infant formula, which has a fixed nutritional profile, human milk is dynamic and can vary between and within feedings, influenced by maternal factors, environmental conditions, and the methods used for expressing and managing milk.

A thorough understanding of human milk's composition is vital for effective infant feeding, especially for at-risk infants. It also allows for the assessment of how milk storage and pasteurization may impact its beneficial components. Additionally, some of the bioactive elements in human milk are being studied for their potential use in medical treatments, either as preventive or therapeutic agents.

4.2 Nutritional Components of Human Milk

Human milk is a complex and dynamic fluid that provides essential nutrition for infants, with its macronutrient composition varying among mothers and throughout lactation. However, the overall nutritional quality of human milk remains relatively stable across different populations, regardless of the mother's nutritional status. On average, mature, full-term human milk contains approximately 0.9 to 1.2 grams per deciliter (g/dL) of protein, 3.2 to 3.6 g/dL of fat, and 6.7 to 7.8 g/dL of lactose, offering an energy content of 65 to 70 kilocalories per deciliter (kcal/dL). The energy content is mainly determined by the fat content of the milk.

Various factors influence the macronutrient composition of human milk, including maternal body weight, protein intake, the number of previous births (parity), the return of menstruation, and the frequency of breastfeeding. While the overall nutritional quality remains consistent, certain nutrients such as vitamins and fatty acids can be influenced by the mother's diet. Therefore, it is important to monitor the mother's nutritional intake to ensure that the infant receives all the necessary nutrients for optimal growth and development.

4.2.1 Macronutrients

Human milk presents a complex composition of proteins, fats, and carbohydrates crucial for infant nutrition and development.

Proteins: It contains a diverse array of proteins and peptides, including secretory immunoglobulin IgA, lysozyme, casein, α-lactalbumin, lactoferrin and serum albumin. These proteins play vital roles in immunity and growth. Notably, the protein content tends to decrease over the first few weeks of lactation, irrespective of delivery timing.

Fat: Rich inoleic and palmitic acids,fat of human milk varies in concentration between foremilk and hindmilk, with hindmilk typically containing higher fat levels. Maternal dietary choices influence the fatty acid composition, especially in omega-3 and omega-6 fatty acids.

Carbohydrates: Lactose predominates as the primary carbohydrate, maintaining relative stability in concentration. However, lactose levels may be higher mothers milk producing larger volumes. Human milk also contains oligosaccharides, important for gut health and immunity, with concentrations varying by lactation stage and maternal genetics.

Understandingcomposition of human milk is vital for optimizing infant nutrition, particularly for infants requiring additional supplementation.

Maternal health and dietary factors significantly influence milk composition, underscoring maternal well-being importance in infant nutrition.

4.2.2 Micronutrients

Human milkrecognized as a standard for infant's nutrition, but its micronutrient composition can vary based on factors such as maternal body stores and diet. Micronutrients like vitamins, B6, B12,A, B1, B2D,iodine may fluctuate accordingly. Given potential dietary insufficiencies, ongoing multivitamin supplementation during lactation is advised.

Notably, human milk tends to be low in vitamin K regardless of maternal diet, while vitamin D levels may also be inadequate, especially with limited maternal sunlight exposure—a common issue globally. While research explores the effects of maternal supplementation with Vitamin D on milk composition, current pediatric guidelines prioritize postnatal vitaminsupplementation for breastfeeding. These measures aim to ensure infants receive essential nutrients crucial for their health and development, compensating for potential maternal dietary limitations.

4.3 Bioactive Components Alongwith Their Sources

Human milk contains a variety of bioactive components that significantly influence biological processes and impact infant health. These components come from different sources: some are produced and secreted by the mammary epithelium, others originate from cells within the milk itself, and some are transported from maternal blood via receptor-mediated processes across the mammary epithelium. Additionally, the secretion of milk fat globules carries a wide range of membrane-bound proteins and lipids, further enriching the bioactive profile of human milk.

For example, during lactation, antigen-specific B cells migrate to the mammary gland, where polymeric immunoglobulin receptors (pIgR) facilitate the transfer of IgA into the milk duct lumen. Likewise, vascular endothelial growth factor (VEGF) is found in much higher concentrations in breast milk than in maternal serum, suggesting it originates from the mammary gland itself.

Understanding the origins of these bioactive components is crucial for explaining the variability in milk composition, especially following maternal intake of specific medications. While a comprehensive examination of all bioactive factors would be beyond the scope of this discussion, we will focus on those factors that vary between mothers of term and preterm infants or change throughout the lactation period. This reflects the remarkable ability of human milk to adapt its composition to meet the unique needs of the infant.

Many of these components interact in a synergistic manner, underscoring the superior benefits of breastfeeding over supplementation with isolated bioactive factors or combinations of those factors.

4.4 Major bioactive Factors in Human Milk

Immunoglobulins name	Function
IgA/sIgA	Pathogen binding inhibition
IgG	Anti-microbial, activates Phagocytosis(IgG1, IgG2, IgG3); anti-inflammatory, response to allergens (IgG4)
IgM	Agglutination, complement activation

Cells name Function	Function
Macrophages	Protecting against infection and T-cell activation
Stem cells	Regeneration with repair

Chemokines names	Function
G-CSF	Trophic factor in intestines
MIF	Macrophage Migratory Inhibitory Factor: Prevents macrophages from movement, Increases the anti-pathogen activity of the macrophages

Anti-microbial names	Functions
Lactoferrin	Acute phase proteins, chelation of iron, antibacterial,anti-oxidant
Lactadherin/MFG E8	Anti-viral, prevents inflammation by enhancing phagocytosis of apoptotic cells

Metabolic hormones names	Functions
Adiponectin	Reduced infant BMI and weight, Anti-inflammatory function
Leptin	Regulates energy conversion and infant BMI, regulation of apetite
Ghrelin	energy conversion and InfantBMI

Oligosaccharides & glycans name	Functions
HMOS (Human Milk Oligosaccharides)	Prebiotic, stimulating beneficial colonization and reducing colonization with pathogens; reduced inflammation
Gangliosides	Brain development; anti-infectious
Glycosaminoglycans	Anti-infectious

Mucin name	Function
MUC1	Prevents infection by bacteria and virus
MUC4	Prevents infection by bacteria and virus

Cytokine Inhibitors	Functions
TNFRI and II	Inhibits TNFα, anti-inflammatory

Hormones name	Functions
Calcitonin	Develops enteric neurons
Somatostatin	Regulates gastric epithelial growth

Growth Factors name	Functions
EGF	Stimulates cell proliferation andmaturation
HB-EGF	Protects against damage fromhypoxiaand ischemia
VEGF	Promotes angiogenesis and tissuerepair
NGF	Promotes neuron growth and maturation
IGF	Stimulates growth anddevelopment,increased RBCs and hemoglobin
Erythropoietin	Erythropoiesis, development of intestine

Cytokines name	Functions
IL-6	Stimulation of the acute phaseresponse, B cell activation, pro-inflammatory
IL-7 Increased thymic size and output	Increased thymic size and output
IL-8 Recruitment of neutrophils, proinflammatory	Recruitment of neutrophils, proinflammatory
IL-10	Repressing Th1-type inflammation, induction of antibody production, facilitation of tolerance
IFNγ	Pro-inflammatory, stimulates Th1 response
TGFβ	Anti-inflammatory, stimulation of T cell phenotype switch
TNFα	Stimulates inflammatory immune activation

4.5 Growth Factors

Human milk contains a variety of growth factors that influence multiple physiological systems, supporting the development and function of the intestinal tract, vasculature, nervous system, and endocrine system. These growth factors have specific roles in the infant's growth and health:

1. **Intestinal Maturation and Repair**: **Epidermal Growth Factor (EGF)** promotes the development and repair of the intestinal lining, stimulating epithelial cell growth in the gastrointestinal tract to ensure proper maturation and healing.
2. **Development of the Enteric Nervous System**: **Neuronal Growth Factors** are essential for the growth and maturation of the enteric nervous system (ENS), which controls gastrointestinal functions like digestion and motility.

3. **Tissue Growth**: The **Insulin-like Growth Factor (IGF) Superfamily** regulates the growth and development of tissues. IGFs are vital for promoting cellular growth and repair, supporting overall tissue development in infants.
4. **Vascular System Regulation**: **Vascular Endothelial Growth Factor (VEGF)** plays a key role in the formation of new blood vessels, a process known as angiogenesis. This helps maintain proper vascular function and ensures that tissues receive adequate oxygen and nutrients.
5. **Intestinal Development and Anemia Prevention**: **Erythropoietin (Epo)** is important for red blood cell production, which helps prevent anemia. It also supports intestinal development by stimulating the growth of gastrointestinal tissues.
6. **Growth-Regulating Hormones**: **Calcitonin** and **Somatostatin** are involved in regulating growth. Calcitonin helps maintain calcium balance and bone health, while somatostatin regulates the release of growth hormone and insulin.
7. **Regulation of Metabolism and Body Composition**: **Adiponectin** and other hormones play a role in controlling metabolism and body composition. These hormones help manage energy balance, fat storage, and metabolic processes, contributing to the infant's overall growth.

These bioactive components in human milk are crucial for supporting the infant's development and health, influencing a wide range of physiological functions that help the infant grow, develop, and thrive.

4.6 Immunological Factors

Feeding human milk protects against infection and inflammation, and early milk is enriched in immune factors that help to ensure infant survival. The specific protective components of human milk are so numerous and multi-functional, that science is just beginning to understand their functions.

1. Transfer of living protection and programming: Cells of human milk
2. Communication between cells: Cytokines and chemokines
3. Protection from infection: Acquired and innate factors
4. Selection for the growth of beneficial organisms: Oligosaccharides

4.7 Human Formula

- It is a manufactured food designed and marketed for feeding to babies under 12 months of age.

- Lower protein and lactose content(easily digestible)
- Higher lactose content
- Its need to be used within 1 month of opening the container

4.8 Types of Infant Formulas

First stage formula: It is also known as infant formula, formulated for newborns and young infants up to approximately six months old. These formulas typically contain a blend of whey protein and casein, mirroring the protein composition of human milk. Whey protein, resembling that of human milk, aids in easy digestion and meets the nutritional needs crucial for early growth and development.

Second Stage Formula: It is also referred to as follow-on formula or stage 2 formula, designed for older infants, typically aged six months and beyond. Unlike first stage formulas, second stage formulas are primarily casein-dominant. Casein, a protein digested more slowly, may be better suited for older babies with more developed digestive systems. Additionally, second stage formulas usually boast higher levels of iron, protein, and other nutrients, meeting the increased nutritional demands of older infants transitioning to solid foods.

4.9 Composition of Infant Food Formulation

- **Protein Source:** Derived from purified cow's milk whey and casein, the protein component of infant food formulation ensures a balanced intake crucial for supporting infant growth and development. These proteins furnish essential amino acids vital for muscle and tissue formation.
- **Fat Source:** An intricately blended combination of vegetable oils is integrated into infant food formulation to provide a well-rounded source of healthy fats. Rich in essential fatty acids such as omega-3 and omega-6, these oils play a pivotal role in fostering brain and nervous system development.
- **Carbohydrate Source:** Lactose, a naturally occurring sugar inherent in milk, serves as the primary carbohydrate source in infant food formulation. Lactose offers readily available energy for growing infants while facilitating the absorption of essential nutrients like calcium.
- **Fortifications:** To ensure comprehensive nutritional support, infant food formulation undergoes fortification with a precise blend of vitamins, minerals, and amino acids. These additions serve to bridge potential nutrient gaps, thereby promoting optimal growth and development during infancy.

- **Additional Ingredients:** Tailored to meet the unique needs of infants, infant food formulation may incorporate other ingredients selected by the manufacturer. These could include natural flavours, prebiotics, probiotics, or dietary fibers, all aimed at enhancing nutritional content, taste, or texture.

In essence, infant food formulation is meticulously crafted to deliver a complete and balanced source of nutrition, encompassing essential proteins, fats, carbohydrates, vitamins, minerals, and fortifications crucial for supporting healthy infant growth and development.

4.10 FSSAI Standard of Infant Formula

Parameters (% by weight)	**Requirements for 100 g**
Moisture (% by weight) (max)	4.50
Total protein	10.00- 16.00
Total fat (min), gram milk fat	18.0- 25.00
a) Linoleicacid, mg b) α-Linolenicacid,mg,Min c) RatioofLinoleicacid/α-Linolenicacid, Min.	1500.00-7000.00 250.00 6:1
Carbohydrates	45.00 - 70.00
Total ash (max)	8.50
Ash insoluble in dilutehydrochloricacid,percentbyweight,Max	0.10
Vitamin A (as retinolequivalent,RE),µg	350.00-823.00
Vitamin D, µg	5.00 – 14.00
Vitamin E, mg	2.50 – 6.00
Vitamin K, µg	7.50-19.00
Vitamin C, mg	25.00 – 75.00
Thiamine,µg	200.00-517.00
Riboflavin,µg	400.00-2000.00
Niacinequivalent,mg	3.80-9.90
Pyridoxine, µg	100.00-400.00
Dietary Folate equivalent(DFE),µg	15.00-56.90
Pantothenicacid,mg	2.00-10.00
VitaminB12,µg	0.25– 4.0
Biotin,µg	7.50-50.00
Choline,mg,Min	32.00
Sodium, mg	90.00 – 300.00
Potassium, mg	300.00 – 900.00
Chloride,mg	250.00-800.00
Calcium,mg	250.00-700.00
Phosphorous,mg	125.00-500.00
Calcium:Phosphorusratio	1:1- 2:1

Parameters (% by weight)	Requirements for 100 g
Magnesium,mg	30.00–75.20
Iron, mg	3.00 – 7.00
Iodine,µg	90.00 – 225.60
Copper,µg	160.00-470.00
Zinc, mg	2.50-5.90
Manganese,µg	5.00-500.00
Selenium,µg	5.0 -17.00

a) Optional Ingredients: When prepared according to the instructions for use, infant formula may include additional nutrients that are naturally present in human milk, in the following specified amounts:

Sl.No.	Nutrients	Requirementsper100mlofthe productreadyforconsumption
1.	Carotenes,mg,Min	0.025
2.	Aminoacids(Lforms),mg,Min	0.90
3.	Non-proteinnitrogen,mg,Min	17.00
4.	Nucleotides,mg,Min	1.17
5.	L-carnitine,mg,Min	0.72
6.	Lactalbumin,mg,Min	140.00

Sl. No.	Nutrients	Requirementsper100mlofthe productreadyforconsumption
7.	Lactoferrin,mg,Min	27.00
8.	Lysozyme,mg,Min	80.00
9.	Glucosamine,mg,Min	70.00
10.	Inositol,mg,Min	2.70
11.	Citricacid,mg,Min	35.00
12.	Cholesterol,mg,Min	8.80
13.	Fucose,mg,Min	130.00
14.	Lipidphosphorous,mg,Min	0.70
15.	Prostaglandins,mg,Min	PGE15.00 PGF40.00
16.	Taurine,mg,Max	8.40
17.	Molybdenum,µg	0.90-6.50
18.	Chromium,µg	1.90 -6.50

b) Foodadditives

The following food additives may be included in the preparation of infant formula for consumption, when made according to the manufacturer's guidelines, unless stated otherwise:

Food Additive	INSNo.	Recommended maximum level per 100ml of the product ready for consumption
Thickeners		
Guargum	412	0.1gminliquidformulascontaining hydrolysedprotein
Carobbeangum (Locustbean gum)	410	0.1gminalltypesofinfantformula
Distarchphosphate	1412	0.5gmsinglyorincombinationinsoybasedinfant formulaonly
Acetylateddistarchphosphate	1414	
Phosphated distarch phosphate	1413	

Food Additive	INSNo.	Recommended maximum level per 100ml of the product ready forconsumption
Hydroxypropylstarch	1440	2.5gmsinglyorincombinationin hydrolysedproteinand/or aminoacidbasedinfant formulaonly
Carrageenan	407	0.03gm (inregularmilk and soybasedliquidinfantformulaonly) 0.1gminhydrolysedprote in and/oramino acid basedliquidinfant formulaonly
Emulsifiers		
Lecithin	322	0.5gminall types of in fantformula*
Mono-anddiglycerides	471	0.4gminall types of infantformula*
Citricandfatty acid esters of glycerol	472c	0.9gminalltypes of liquidinfant formula 0.75gminalltypes of powdered infantformula
Acidity Regulators		
Sodiumhydroxide	524	0.2 gm singly orin combination and with in the limits for sodium, potassiumandcalciuminprovision (g) of sub-regulation7(1) in all typesofinfant formula
Sodiumhydrogencarbonate	500ii	
Sodiumcarbonate	500i	
Potassiumhydrogencarbonate	501ii	
Potassiumcar bonate	501i	
Potassiumhy droxide	525	
Calciumhy droxide	526	
L(+)lacticacid	270	GMPinalltypes of in fantformula
Citricacid	330	
Sodiumdihydrogencitrate	331i	
Trisodiumcitrate	331iii	
Potassiumcitrate	332	

FoodAdditive	INSNo.	Recommended maximum levelper 100 ml of the productready for consumption
Sodium dihydrogen phosphate, disodium hydrogen phosphate and trisodium phosphate	339i,iiandiii	45 mg as phosphorous singly or incombination and with in limits for sodium, potassium and phosphorous
Potassium dihydrogen phosphate, dipotassium hydrogen phosphate and tripotassium phosphate	340i,iiandiii	
Antioxidants		
Mixedtocopherolconcentrate	307b	1mginalltypes of in fant for mula singlyorincombination
Ascorbylpalmitate	304i	1mginalltype of in fant for mula singlyorincombination
Packaginggases		
Carbondioxide	290	GMP
Nitrogen	941	

4.11 FSSAI Standard for Milk-Cereal Based Complementary Food: This standard is applicable to milk-cereal based complementary foods in the form of powder, small granules, or flakes, intended to supplement the diet of infants from six months to twenty-four months of age.

1. **Composition:** The milk-cereal based complementary food must adhere to the following criteria:
2. **Nutritional Standards**: The food should provide adequate nutrition, ensuring that it contains the necessary amounts of proteins, fats, carbohydrates, vitamins, and minerals essential for the proper growth and development of infants and young children, in accordance with the prescribed regulations.
3. **Safety Compliance**: The food must be free from harmful substances, such as contaminants, pathogens, or toxins, that could pose a risk to the health of the target age group.
4. **Hygienic Production**: The food should be manufactured in sanitary conditions, utilizing appropriate processing methods to ensure safety and maintain quality while preventing contamination.
5. **Additives and Preservatives**: Any additives, preservatives, or colorants used must comply with the approved limits specified by the regulations. Only safe, authorized substances are permitted.
6. **Proper Labeling and Packaging**: The product must be clearly labeled with accurate information regarding the ingredients, nutritional content, age suitability, and other relevant details to guide consumers in making informed decisions.

Prohibited Ingredients: The food must not contain any prohibited ingredients, such as hydrogenated fats or oils, as outlined in the regulatory standards.

These guidelines ensure that milk-cereal based complementary foods are both safe and nutritionally suitable for young children, while complying with safety and quality regulations.

Sl.No.	Parameters	Requirements per 100g
1.	Moisture, percent by weight, Max	5.00
2.	Total protein including milk protein, per cent by weight (N x6.25), Min Milk protein (Nx6.38), percent by weight, Min	15.00 5.00
3.	Total fat (including milk fat), percent by weight, Min Milk fat, percent by weight, Min	7.50 5.00
4.	Total carbohydrates, percentby weight, Min	55.00
5.	Totalash, percent by weight, Max	5.00
6.	Ash insoluble in dilutehydrochloric acid; percentby weight, Max	0.10

Sl.No.	Parameters	Requirements per 100g
7.	Crudefibre(ondrybasis), per cent by weight, Max	1.00
8.	Vitamin A(asretinolequivalent, RE), μg	350.00-823.00
9.	VitaminD, μg	5.00–14.00
10.	VitaminC, mg	25.00-75.00
11.	Thiamine, mg	0.20-0.50
12.	Riboflavin, mg	0.40-2.00
13.	Niacinequivalent, mg	3.80-9.90
14.	DietaryFolateequivalent(DFE)*, μg	15.00-50.00
15.	Iron, mg	3.00-7.00
16.	Zinc, mg	2.50-5.90

*1microgram DFE=0.6 microgramfolicacid

a) Optional Ingredient or Nutrient: It may also contain optional ingredient or nutrient as below:

Sl.No.	Nutrient	Requirements per 100g
1.	Pantothenicacid,mg	2.00-10.00
2.	VitaminB12,mg	0.25-0.70
3.	VitaminK,mg	7.50-19.00
4.	Choline,mg,Max	32.00
5.	Inositol,mg	20.00 200.00
6.	Calcium,mg	250.00-700.00
7.	Phosphorus,mg	125.00-500.00
8.	Chloride,mg	250.00-800.00

Sl.No.	Nutrient	Requirements per 100g
9.	Magnesium,mg	30.00-75.20
10.	Sodium,mg	90.00- 300.00
11.	Selenium,mg	5.00-17.00
12.	Taurine,mg,Max	60.00
13.	L-aminoacids,mg,Min	0.90
14.	L-Carnitine,mg,Min	5.00
15.	Biotin,µg	7.50-50.00
16.	Iodine,µg	90.00-225.60
17.	Potassium, mg	300.00-900.00
18.	Pyridoxine,µg	100.00-400.00

b) Food Additives

Givenadditives are permitted in the preparation of milk cereal based complementary food for in fants in 100 gm of the product ready for consumption prepared following manufacturer' sinstructions, unless other wise in dicated.

Food Additive	INSNo.	Recommended maximum Level per 100g of the product ready for consumption
Emulsifiers		
Lecithins	322	1500mg
Mono-anddiglycerides	471	5000mg singlyorin combination
Acetic glycerol fatty acid esters of	472a	
Lactic glycerol fatty acid esters of	472b	
Citricacidfatty acid esters of glycerol	472c	
Acidityregulators		
Sodiumhydrogencarbonate	500ii	
Potassiumhydrogencarbonate	501ii	
Calciumcarbonate	170i	

Food Additive	INSNo.	Recommended maximum Level per 100g of the product ready for consumption
L(+)lacticacid	270	GMP
Citricacid	330	
Aceticacid	260	
Potassiumacetate	261	
Sodiumacetate	262i	
Calciumacetate	263	
Malicacid(DL)-L(+)formonly	296	
Sodiumlactate(solution)-L(+) formonly	325	
Potassium lactate (solution)- L(+)formonly	326	
Calciumlactate-L(+)formonly	327	
Monosodiumcitrate	331i	

Food Additive	INSNo.	Recommended maximum Level per 100g of the product ready for consumption
Trisodiumcitrate	331ii	
Monopotassiumcitrate	332i	
Tripotassiumcitrate	332ii	
Calciumcitrate	333	
Hydrochloricacid	507	
Sodiumhydroxide	524	
Potassiumhydroxide	525	
Calciumhydroxide	526	
Glucono-delta-lactone	575	GMP
L(+)Tartaricacid	334	500mgsinglyorin combination
Disodiumtartarate	335ii	
PotassiumsodiumL(+)tartrate L(+)formonly	337	
Orthophosphoricacid	338	OnlyforpHadjustment
Monosodiumorthophosphate	339i	
Disodiumorthophosphate	339ii	
Trisodiumorthophosphate	339iii	

Food Additive	INSNo.	Recommended maximum Level per 100g of the product ready for consumption
Monopotassium or thophosphate	340i	440mg singlyorincombinati on asphosphorous
Dipotassiumorthophosphate	340ii	
Tripotassiumorthophosphate	340iii	
Monocalciumorthophosphate	341i	
Dicalciumorthophosphate	341ii	
Tricalciumorthophosphate	341iii	
Antioxidants		
Mixedtocopherolconcentrate	306	300mg/kg fatoroilsinglyorincombination
Alphatocopherol	307	
L-ascorbylpalmitate	304	200mg/kgfatoroil
L-ascorbicacid	300	50mgexpressedasascorbicacid
Sodiumascorbate	301	
Potassiumascorbate	303	
Calciumascorbate	302	20mgexpressedasascorbic acid
Raisingagent		
Ammoniumcarbonate	503i	GMP
Ammonium hydrogen carbonate	503ii	
Sodiumcarbonate	500i	
sodiumhydrogencarbonate	500ii	

Food Additive	INSNo.	Recommended maximum Level per 100g of the product ready for consumption
Thickeners		
Carobbeangum	410	1000mgsinglyorin combination 2000 mg in gluten free cereal basedfoods
Guargum	412	
GumArabic	414	
Xanthangum	415	
Pectins (amidated and non amidated)	440	

Food Additive	INSNo.	Recommended maximum Level per 100g of the product ready for consumption
Oxidizedstarch	1404	5000mg singlyorin combination
Monostarchphosphate	1410	
Distarchphosphate	1412	
Phosphateddistarchphosphate	1413	
Acetylateddistarchphosphate	1414	
Acetylateddistarchadipate	1422	
Starchacetateesterifiedwith aceticanhydride	1420	
Starchsodiumoctenylsuccinate	1450	
Acetylatedoxidizedstarch	1451	
Anticakingagent		
Silicondioxide	551	GMP
Packaginggases		
Carbondioxide	290	GMP
Nitrogen	941	
Enzymes		
Alphaamylase	-	GMP

4.12 FSSAI Standard of Processed Cereal Based Complementary Food

This standard applies to processedcereal based complementary food in powder, small granules or flakes form intended to complement the diet of infant after six months upto twenty-four months of age.

Composition

The processed cereal-based complementary food must fulfill the following criteria:

1. **Nutritional Balance**: The product should provide a well-rounded nutritional profile, offering the necessary energy and nutrients to support the growth and development of infants and young children, following recognized dietary standards.

2. **Ingredient Proportions**: The food must contain at least 75% milled cereals and legumes, which may include pulses, soybeans, millets, and other ingredients like nuts and edible oil seeds.
3. **Sodium Content**: The sodium content in the product must be limited to a maximum of 100 mg per 100 kcals of the ready-to-eat serving to ensure it is within acceptable limits for young children.
4. **Restricted Ingredients**: The product must not contain hydrogenated vegetable oils or fats, ensuring it is free from unhealthy trans fats.
5. **Safe Processing**: If honey or maple syrup is included, it must undergo processing that effectively eliminates Clostridium botulinum spores, preventing the risk of botulism.
6. **Vitamins and Minerals**: Any added vitamins and minerals should comply with the Recommended Dietary Allowances (RDA) set by the Indian Council of Medical Research (ICMR). If no RDA is provided, the Codex Alimentarius Commission's standards should apply. The levels of these nutrients must not exceed the recommended amounts.
7. **Quality and Safety Compliance:** The product must adhere to necessary safety, nutritional, and manufacturing standards, ensuring it is free from contaminants and safe for consumption by infants and young children.

Sl.No.	Parameters	Requirements per 100g
1.	Moisture, percent by weight, Max	5.00
2.	Total protein (Nx6.25), percent by weight, Min	15.00
3.	Total fat, percent by weight, Max	7.50
4.	Total carbohydrates, percent by weight, Min	55.00
5.	Totalash, percent by weight, Max	5.00
6.	Ash insoluble in dilutehydrochloric acid, percentby weight, Max	0.10
7.	Crudefibre (ondrybasis), per cent by weight, Max	1.00
8.	VitaminA(asretinolequivalent), µg	350.00-823.00
9.	VitaminD,µg	5.00-14.00
10.	VitaminC,mg	25.00-75.00
11.	Thiamine,mg	0.20-0.50
12.	Riboflavin,mg	0.40-2.00
13.	Niacinequivalent,mg	3.80-9.90
14.	Dietary Folateequivalent(DFE), µg	15.00- 50.00
15.	Iron,mg	3.00-5.65
16.	Zinc, mg	2.50-5.90
17.	Pantothenicacid,mg	2.00-10.00
18.	Pyridoxine,µg	100.00-400.00
19.	VitaminB12,µg	0.25-0.70

Sl.No.	Parameters	Requirements per 100g
20.	Biotin,µg	7.50-50.0
21.	Choline,mg,Min	32.00
22.	Inositol,mg	20.00- 200.00
23.	Selenium,µg	5.00-17.00

1microgramDFE= 0.6microgramfolicacid

b) Food Additives

The following additives are allowed in the preparation of processed cereal-based complementary food for infants in 100 g of the product ready for consumption, prepared according to the manufacturer's instructions, unless otherwise stated:

1. **Vitamins and Minerals**: Specific vitamins and minerals can be added to improve nutritional value, provided the amounts align with the Recommended Dietary Allowances (RDA) set by the Indian Council of Medical Research (ICMR). If RDA values are not available, the guidelines set by the Codex Alimentarius Commission should be followed.
2. **Proteins and Amino Acids**: Protein concentrates, isolates, and amino acids may be included to boost protein content, ensuring the product fulfills the nutritional needs of infants.
3. **Carbohydrates**: Ingredients like malt, corn syrup, or other similar carbohydrates may be used to provide energy, ensuring they meet safety and nutritional standards.
4. **Edible Oils and Fats**: Only edible vegetable oils, defatted edible oil seed extracts, and fats are permitted. The use of hydrogenated vegetable oils and fats is strictly prohibited.
5. **Emulsifiers and Stabilizers**: Safe food-grade emulsifiers and stabilizers may be used, provided they comply with the acceptable limits set by regulatory authorities for infant safety.
6. **Antioxidants**: Antioxidants approved by regulatory authorities may be included to preserve the product, as long as they remain within the allowed limits.
7. **Flavoring Agents**: Safe, approved flavoring agents or sweeteners, such as honey or maple syrup, may be used. However, if honey is included, it must be processed to eliminate *Clostridium botulinum* spores to prevent botulism.

8. **Preservatives**: Preservatives may only be used if necessary to ensure the product's safety and stability, adhering to applicable safety regulations.

Food Additive	**INSNo.**	**Recommended maximum levelper 100g of the product ready for consumption**
Emulsifiers		
Lecithins	322	1500mg
Mono-anddiglycerides	471	5000mgsinglyorin combination
Aceticfattyacidestersof glycerol	472a	
Lacticfattyacidesters of glycerol	472b	
Citricacidfattyacidestersof glycerol	472c	
Acidityregulators		
Sodiumhydrogencarbonate	500ii	GMP
Potassium hydrogen carbonate	501ii	
Calciumcarbonate	170i	
L(+)lacticacid	270	
Citricacid	330	
Aceticacid	260	
Potassiumacetate	261	
Sodiumacetate	262i	
Calciumacetate	263	
Malicacid (DL)-L(+) form only	296	
Sodiumlactate (solution)- L(+) formonly	325	
Potassium lactate (solution)-L(+) formonly	326	
Calciumlactate-L(+) form only	327	
Monosodiumcitrate	331i	
Trisodiumcitrate	331ii	
Monopotassiumcitrate	332i	
Tripotassiumcitrate	332ii	
Calciumcitrate	333	
Hydrochloricacid	507	
Sodiumhydroxide	524	
Potassiumhydroxide	525	
Calciumhydroxide	526	
Glucono-delta-lactone	575	GMP
L(+)Tartaricacid	334	500mgsinglyorin combination
Disodiumtartarate	335ii	
Potassium sodium L(+) tartrateL(+) formonly	337	

Orthophosphoricacid	338	Onlyforp Hadjustment 440mgsinglyorincom binationasphosphorous
Mono sodium ortho phosphate	339i	
Disodiumorthophosphate	339ii	
Trisodiumorthophosphate	339iii	
Monopotassium orthophosphate	340i	
Dipotassium orthophosphate	340ii	
Tripotassium orthophosphate	340iii	
Monocalcium orthophosphate	341i	
Dicalciumorthophosphate	341ii	
Tricalciumorthophosphate	341iii	
Antioxidants		
Mixed tocopherol concentrate	306	300mg/kg fat oroil singlyorin combination
Alphatocopherol	307	
L-ascorbylpalmitate	304	200mg/kgfatoroil
L-ascorbicacid	300	50mgexpressedasascorbic acid
Sodiumascorbate	301	
Potassiumascorbate	303	
Calciumascorbate	302	20mgexpressedasascorbic acid
Raisingagent		
Ammoniumcarbonate	503i	GMP
Ammonium hydrogen carbonate	503ii	
Sodiumcarbonate	500i	
sodiumhydrogencarbonate	500ii	
Thickeners		
Carobbeangum	410	1000mgsinglyorin combination 2000mgingluten freecerealbasedfoods
Guargum	412	
GumArabic	414	
Xanthangum	415	
Pectins(amidatedandnon amidated)	440	
Oxidizedstarch	1404	5000mgsinglyorin combination
Monostarchphosphate	1410	
Distarchphosphate	1412	
Phosphateddistarch phosphate	1413	
Acetylated distarch phosphate	1414	
Acetylateddistarchadipate	1422	
Starch acetate esterified withaceticanhydride	1420	
Starch sodium octenyl succinate	1450	
Acetylatedoxidizedstarch	1451	
Anticakingagent		
Silicondioxide	551	GMP

Packaginggases		
Carbondioxide	290	GMP
Nitrogen	941	
Enzymes		
Alphaamylase	-	GMP

4.13 Cow Milk vs Human milk

Unsuitability of Cow's Milk for Infants: Plain cow's milk is not appropriate for infants due to its high casein and low whey content, which differs significantly from human breast milk's composition.

Casein to Whey Protein Ratio in Cow Milk: Cow milk typically boasts a casein to whey protein ratio of roughly 80:20, a considerable variance from the ratio found in human milk.

Casein to Whey Protein Ratio in Infant Milk: Infant milk formulations undergo processing to adjust the protein composition to closely resemble that of human milk. Early lactation formulations exhibit a ratio around 10:90, gradually transitioning to 40:60 in mature milk formulations, and eventually settling at a balanced 50:50 ratio in late lactation formulations.

Recommendation Regarding Untreated Cow's Milk:Due to its unsuitability for infants, untreated cow's milk is not recommended for consumption before the age of 12 months. This caution stems from potential digestive challenges arising from the high casein content and imbalanced protein composition compared to human milk.

Infant Formulation Processing: Infant milk formulations undergo meticulous processing to enhance protein digestibility and achieve a whey to casein protein balance akin to human milk. This processing ensures that the nutritional profile of infant formula aligns closely with the needs of growing infants, facilitating optimal growth and development.

In essence, the incompatibility of plain cow's milk for infants underscores the necessity of specialized infant milk formulations, which undergo processing to mirror human milk's protein composition. This ensures enhanced digestibility and provision of essential nutrients vital for healthy infant development.

4.14 Special Types of infant formulas

1. Formulas for Protein Allergic Babies

Partially Hydrolyzed Formulas:These formulas feature proteins that have undergone partial hydrolysis, breaking them into smaller fragments. They aim to ease digestion for infants with mild protein allergies, potentially reducing allergic reactions.

Extensively Hydrolyzed Formulas (Hypoallergenic): These formulas contain proteins extensively broken down into tiny fragments, making them highly hypoallergenic. They are suitable for infants with severe protein allergies who cannot tolerate partially hydrolyzed formulas.

Amino Acid-Based Formulas: These formulas contain individual amino acids instead of whole proteins, making them the least likely to cause allergic reactions. While effective for infants with severe allergies, they tend to be more expensive due to their specialized composition.

2. Lactose-Free Infant Formulas

Designed for Lactose Intolerance: Specifically formulated for infants aged 0-12 months with lactose intolerance, these formulas cater to those unable to digest lactose, the sugar found in milk.

Cow's Milk Protein Based: Despite being lactose-free, these formulas still utilize cow's milk protein. However, they undergo processing to eliminate lactose, rendering them suitable for infants with lactose intolerance.

Alternative Carbohydrate Sources: Instead of lactose, lactose-free infant formulas incorporate alternative carbohydrate sources such as glucose, sucrose, natural sugars, or modified starches. These alternatives provide necessary energy for infant growth and development without triggering digestive issues in lactose-intolerant babies.

In summary, formulas for protein allergic babies offer a range of options tailored to allergy severity, while lactose-free infant formulas provide suitable nutrition for infants with lactose intolerance through alternative carbohydrate sources, ensuring their nutritional needs are met without compromising digestive comfort.

5

Practical Protocols of Concentrated and Dried Milks

Experiment No.- 1

AIM

To determine the sucrose content in condensed or evaporated milk using the Lane-Eynon method.

Principle

The Lane-Eynon method is used to measure sucrose in condensed milk by utilizing the reducing properties of sugars. Sucrose, being a non-reducing sugar, is first hydrolyzed into glucose and fructose (reducing sugars). These reducing sugars then react with Fehling's solution, producing a brick-red precipitate of cuprous oxide (Cu2O) when heated. This allows for the accurate determination of sucrose content after the hydrolysis (inversion) process.

Reactions

- **Reaction of Fehling's Solution**
  - $CuSO_4 + NaOH \rightarrow Cu(OH)_2 + Na_2SO_4$
  - $Cu(OH)_2$ + Reducing Sugar (RCHO) $\rightarrow$ Cu_2O (brick-red precipitate) + RCOOH

This reaction forms cuprous oxide (Cu_2O), a brick-red precipitate. The intensity of this precipitate correlates with the concentration of reducing sugar in the sample, providing a way to determine sucrose content.

Reagents

1. **0.1N Sodium Hydroxide Solution**: Prepare by dissolving 4g of NaOH in 1 liter of water.
2. **Stock Solution of Invert Sugar**: Dissolve 23.75g of sucrose in 100mL of distilled water. Add 10mL of concentrated hydrochloric acid (HCl) and mix. Allow it to sit for 3 days and adjust the volume to 1 liter.

3. **Standard Solution of Invert Sugar**: Neutralize the stock invert sugar solution with NaOH and verify with litmus paper. Dilute 50mL of this neutralized solution to 500mL.
4. **Fehling A Solution**: Dissolve 34.639g of copper sulfate pentahydrate in water, add 0.5mL of concentrated sulfuric acid, and dilute to 500mL.
5. **Fehling B Solution**: Dissolve 173g of Rochelle salt and 50g of NaOH in water. After 2 days, filter and dilute to 500mL.
6. **2N Zinc Acetate Solution**: Dissolve 21.9g of zinc acetate in water and add 3mL of glacial acetic acid. Dilute to 100mL.
7. **1N Potassium Ferrocyanide Solution**: Dissolve 10.6g of potassium ferrocyanide in water and dilute to 100mL.
8. **Concentrated Hydrochloric Acid (HCl)**: This solution has a density of 1.16g/mL.
9. **Concentrated Ammonia Solution**: Density of 0.88g/mL.
10. **Dilute Ammonia Solution (10%)**: Dilute 10mL of concentrated ammonia solution to 100mL with water.
11. **0.2% Methylene Blue Indicator**: Dissolve 0.2g of methylene blue in 100mL of water.
12. **Dilute Acetic Acid Solution (10%)**: Dilute 10mL of acetic acid to 100mL with water.

Rapid Procedure for Inversion

1. Dissolve 11.88g of sucrose in 75mL of distilled water in a 500mL flask.
2. Slowly add 10mL of 6.34N hydrochloric acid (HCl) while gently rotating the flask.
3. Heat the solution in a water bath at 70°C for 5 minutes, then cool to 20°C.
4. Neutralize the solution with 1N sodium hydroxide and adjust the volume to 500mL with distilled water. This hydrolysis process converts sucrose into glucose and fructose.

Glassware and Apparatus

- Water bath
- Burette
- Pipettes (2mL, 10mL)
- 100mL beaker
- Glass rod
- Stopwatch

Procedure

A. Standardization of Fehling's Solution Against Standard Invert Sugar Solution (2.5 mg/mL)

1. Pour the standard invert sugar solution into a 50mL burette.
2. In a 150mL conical flask, mix 5mL of Fehling A and 5mL of Fehling B solutions, then heat the mixture.
3. Titrate the heated mixture with the standard invert sugar solution, continuing the titration until a brick-red precipitate forms.
4. Add 1mL of methylene blue indicator to turn the solution blue.
5. Continue the titration, ensuring the brick-red color reappears. Record the volume of sugar solution used as **V1**.

B. Preparation of Protein- and Fat-Free Sample Filtrate

1. Weigh 40g of well-mixed sample and add 50mL of hot water (80-90°C).
2. Transfer the mixture to a 250mL volumetric flask, then rinse with warm distilled water (60°C) until the total volume reaches 120-150mL.
3. Allow it to cool and add 5mL of 10% ammonia solution.
4. Mix thoroughly and allow it to stand for 15 minutes, then neutralize with dilute acetic acid.
5. To precipitate proteins, add 12.5mL each of zinc acetate and potassium ferrocyanide solutions.
6. Mix, adjust the final volume to 250mL, and filter using Whatman No. 1 filter paper. Label the filtrate as A-1.

C. Estimation of Original Reducing Sugar (Lactose) in the Sample

1. Dilute 25mL of filtrate A-1 to 100mL.
2. Titrate the diluted filtrate with Fehling's solution as in step A, and record the volume of sugar solution used as V2.

D. Estimation of Total Reducing Sugar after Inversion (Glucose + Fructose + Lactose)

1. Transfer 25mL of filtrate A-1 to a 100mL flask. Add 5mL of concentrated hydrochloric acid and incubate at 68°C for 5 minutes.
2. After cooling, neutralize with 1N sodium hydroxide and dilute to 100mL. Label the solution as **B-1**.
3. Titrate solution B-1 with Fehling's solution as described in step A. Record the volume used as **V3**.

Observation

- **V1** = ______ mL (Standard sugar solution used for titration)
- **V2** = ______ mL (Sample solution before inversion)
- **V3** = ______ mL (Sample solution after inversion)

Calculation

Now we know that;

1.10 ml of Fehling (A + B) solution mixture

= V_1 ml of std. sugar solution of C1 mg/ml conc.

(2.5mg / ml)

2. 10 ml of Fehling (A + B) solution mixture = V_2 ml of sample solution before inversion of C_2 mg/ml conc. (Original reducing sugar i.e. lactose)

3. 10 ml of Fehling (A + B) solution mixture= V_3 ml of sample solution after inversion of C_3 mg/ml conc.

(Lactose+Glucose+Fructose)

Now:

$V_1C_1 = V_2C_2 = V_3C_3$

Thus

1) $V_1C_1 = V_2C_2$

   C_2 (mg/ml) = $(V_1C_1) / V_2$

   $$\% C_2 = \frac{(C_1 * V_1 * 250 * 100)}{(V2 * 1000 * 40 * 25)} X\ 100$$

2) $V_1C_1 = V_3C_3$

   C_3 (mg/ml) = V_1C_1 / V_3

   $$\% C_3 = \frac{(C_1 * V_1 * 250 * 100)}{(V_3 * 1000 * 40 * 25)} X\ 100$$

3) % Sucrose= $(C_3 - C_2) * 0.95$

Result: The sucrose content in the condensed milk sample is ______%

Note:

a) 0.95: i.e. on inversion there is slightly higher reading because molecular weight of glucose + fructose = 360 as compared to 342 of sucrose.

Mol. Wt. of Sucrose = 342 = 0.95

Mol. Wt. of glu + fru 360

b) Inversion of sucrose involves breaking down the disaccharide molecule into its constituent monosaccharides, glucose and fructose. This process can be achieved enzymatically with invertase or chemically by boiling with dilute acids. Sucrose, naturally dextrorotatory, rotates plane-polarized light to the right, whereas the resulting mixture of glucose and fructose after inversion is laboratory, rotating light to the left. This change in optical rotation is due to the specific optical properties of glucose and fructose. Inversion of sucrose is crucial in industrial applications, such as in the production of invert sugar, which is sweeter than sucrose and used widely in food processing.

Experiment No. -2

AIM: To determine the sucrose concentration in condensed or evaporated milk using the polarimetric method.

Principle

The polarimetric method for determining sucrose in condensed milk utilizes the Clerget inversion process. This technique selectively hydrolyzes sucrose into glucose and fructose, without affecting other sugars such as lactose. The sample is first treated with ammonia, creating an alkaline medium that stabilizes the solution. It is then neutralized to bring the solution to a neutral pH. Zinc acetate and potassium ferrocyanide are subsequently added to remove impurities, clarifying the solution for polarimetric analysis.

During the process, a sample of the clarified filtrate is first measured for optical rotation before inversion. Sucrose is then hydrolyzed using the Clerget inversion technique. After hydrolysis, the optical rotation is measured again. The difference in optical rotation before and after the inversion allows for calculating the sucrose content. This method is tailored for accurately measuring sucrose in milk products, ensuring selective hydrolysis of sucrose while leaving other sugars unaffected.

Reagents

1. **Zinc Acetate Solution**: Dissolve 21.9g of zinc acetate dihydrate and 3mL of acetic acid in distilled water to prepare 100mL of the solution.
2. **Potassium Ferrocyanide Solution**: Dissolve 10.6g of potassium ferrocyanide in distilled water to prepare 100mL.

3. **Hydrochloric Acid (6.35N ± 0.2N)**: Prepare by diluting concentrated hydrochloric acid to obtain a 6.35 normal (6.35N) concentration, with a tolerance of ± 0.2N.
4. **Ammonia Solution (2 ± 0.2N)**: Prepare by diluting concentrated ammonia to a 2 normal (2N) concentration, with a tolerance of ± 0.2N.
5. **Acetic Acid (10%)**: Mix 10mL of glacial acetic acid with distilled water to prepare 100mL of a 10% acetic acid solution.

Glassware and Apparatus

1. **Polarimeter**: Used to measure the optical rotation of a substance in a liquid sample.
2. **Calibrated Tubes (at least 2 dm in length)**: Used for holding the sample during optical rotation measurements. These tubes are calibrated for precise volume measurements.
3. **Accurately Calibrated Flasks and Pipettes**: Essential for precise volume measurements.
4. **Standard Thermometer (0.1°C accuracy)**: For accurate temperature control during the process, ensuring proper hydrolysis conditions.

Procedure

A. Sample Preparation

1. **Mixing the Sample**: Thoroughly mix the sample to ensure uniformity by using a spoon to rotate it up and down, avoiding air bubbles or froth formation.
2. **Heating (if the sample is old)**: If the sample has aged, heat it to 40°C in a water bath and mix well before proceeding with the analysis.

B. Determining Total Sugar

1. Preparing the Sample for Analysis

- Weigh out 40 grams of the well-mixed sample and place it in a 100mL beaker.
- Add 50mL of hot distilled water (80-90°C) and mix thoroughly.
- Transfer the mixture to a 250mL volumetric flask, rinsing the beaker with hot water at 60°C to ensure all material is transferred, achieving a final volume between 120-150mL in the flask.
- Allow the solution to cool to room temperature, then add 5mL of dilute ammonia solution.

2. Neutralization and Precipitation

- After 15 minutes, neutralize the ammonia by adding dilute acetic acid. The amount required for neutralization can be determined by prior titration.
- Mix the solution thoroughly.
- Add 12.5mL of zinc acetate solution and 12.5mL of potassium ferrocyanide solution in sequence, ensuring gentle mixing after each addition.
- Adjust the volume to the 250mL mark with distilled water and allow the precipitate to settle.

3. Filtration

- Gently mix the solution to avoid air bubbles. If bubbles are present, rotate the flask on a vacuum pump to eliminate them.
- Filter the solution through dry filter paper and discard the first few milliliters of the filtrate, labeling it as A-1.

C. Polarimetric Measurement

- Use the polarimeter to measure the optical rotation of the filtrate at 20°C. Record this as the polarization reading before inversion.

D. Inversion Process

1. Preparation of Inverted Sugar Solution

- Take 50mL of the filtrate **A-1** and add 5mL of concentrated hydrochloric acid (HCl).
- Heat the mixture in a water bath at 68°C for 5 minutes.

2. Neutralization

- After incubation, neutralize the solution with sodium hydroxide (NaOH) solution. Use litmus paper to monitor the pH until it reaches neutrality (pH 7).
- Mix well after each addition of NaOH.

3. Volume Adjustment

- Once neutralized, adjust the volume to the 100mL mark with distilled water.

4. Polarimetric Measurement

- Allow the solution to cool to room temperature (20°C) and then measure its optical rotation using the polarimeter. Record the reading as the polarization after inversion.

Observations

1. Polarization Reading Before Inversion: ______
2. Polarization Reading After Inversion: ______

Calculation

v = (W/100) * (F * 1.08) + (P * 1.55) ....Formula 1

and percentage of sucrose in the sample

%Sucrose= [D - (5/4 * I)]/Q * [(V - v)/v] * [v/(1*W)] ....Formula 2

Where,

W = Weight of the sample taken in grams.

F = Percentage of fat in the sample.

P = Percentage of protein (Nx6.38) in the sample.

V = Volume to which the sample is diluted before filtration.

v = Correction in ml for volume of precipitate produced during clarification.

D = Observed direct polarimeter reading.

I = Observed invert polarimeter reading.

l = Length in dm of polarimeter tube

Q = Inversion divisor factor

Result

The sucrose content in the given condensed milk sample _______.

Note

The value of Q in the equation for optical rotation depends on several critical factors including the type of polarimeter used, the source of light employed, the temperature of the solution (typically measured at 20°C), the concentration of sugars within the solution, and the influence of salts and acids present. These parameters collectively affect the degree of rotation of polarized light passing through the solution, which is fundamental in determining the composition and concentration of sugars, particularly in applications such as analyzing sweetened condensed milk. Calibration and standardization of these measurements are essential to ensure accurate and consistent results in scientific analysis and quality control processes.

a- Sodium lamp= 0.8825

b- Mercury green line (prism or special Waratten screen, 77A)=1.0392

c-International sugar scale (J) light= 2.549

Experiment No. -3

AIM

To determine the fat content in condensed or evaporated milk.

Principle

The fat content in milk can be measured using two primary methods: the **Gerber method** and the **Rose-Gottlieb method**.

- **Gerber Method**: This method is based on volumetric analysis and is widely used due to its simplicity and ability to quickly estimate fat content. It involves mixing milk with sulfuric acid and iso-amyl alcohol in a special Gerber tube. The mixture is then centrifuged to separate the fat, which rises to the top of the tube. The fat percentage is determined by measuring the level to which the fat rises in the calibrated tube.
- **Rose-Gottlieb Method**: This is a more accurate gravimetric technique, which involves breaking the fat globules using ammonia, followed by solvent extraction. The fat is then dried and weighed. This method is more precise but requires more time and careful handling of solvents.

Apparatus

1. **Gerber Butyrometer**: A calibrated tube used to measure fat percentage, available in 6%, 8%, and 10% scales.
2. **Milk Pipette**: With a capacity of 10.75 ± 0.03 mL, suitable for milk at 27°C.
3. **Automatic Dispensers**: For accurately dispensing 10 ± 0.25 mL of sulfuric acid and 1 ± 0.05 mL of iso-amyl alcohol.
4. **Lock Stoppers and Key**: To secure the Gerber butyrometer tubes.
5. **Water Bath**: Used to maintain the temperature of the sample at 27°C during preparation.

6. **Gerber Centrifuge**: For separating the fat from the milk after mixing.
7. **Fat Extraction Flask or Extraction Tube**: For the solvent extraction process in the Rose-Gottlieb method.
8. **Stopper**: Made from cork or synthetic rubber, resistant to fat solvents.
9. **Evaporation Flask or Suitable Container**: Used for evaporating the solvent during the fat extraction process.
10. **Hot Air Oven**: To dry the fat residues after extraction.

Reagents

1. **Ammonia Solution**: With a specific gravity of 0.91.
2. **95% Ethyl Alcohol**.
3. **Peroxide-Free Diethyl Ether**.
4. **Petroleum Ether**: With a boiling range of 40-60°C.
5. **Concentrated Sulfuric Acid**: Approximately 36.8 N.
6. **Iso-Amyl Alcohol**: Furfural-free with a density of 0.807–0.809 g/mL at 27°C.

Preparation of Reagents

To prepare the Gerber sulfuric acid, mix concentrated sulfuric acid with water in a Pyrex flask. Use approximately 900 mL of sulfuric acid to 100 mL of water, and perform the mixing slowly, keeping the mixture cool by placing the flask in an ice-water bath. Stir gently and gradually add the sulfuric acid. After the solution has cooled, measure the specific gravity. Adjust it to the range of 1.807–1.812 g/mL at 27°C. If needed, adjust the gravity by adding water or more acid, always following the same cautious procedure. Store the prepared acid in a glass-stoppered bottle to prevent absorption of moisture.

Procedure

Method 1: Gerber Method

1. Prepare the Sample

- Use an automatic dispenser to transfer 10 mL of sulfuric acid into a Gerber tube. Keep the neck of the tube dry.
- Thoroughly mix the milk sample and fill the pipette slightly above the graduation mark.
- Wipe the pipette and let the milk level settle to the correct mark.
- Release the milk into the Gerber tube, allowing it to drain for three seconds.

2. Add Iso-Amyl Alcohol

- Add 1 mL of iso-amyl alcohol to the tube, seal it with a lock stopper, and shake vigorously to ensure thorough mixing.

3. Centrifugation and Heating

- Place the sealed tube in a water bath set at 65±2°C for 3 minutes to dissolve the casein.
- Centrifuge the tube at 1100 rpm for 5 minutes.
- Ensure the tubes are placed symmetrically in the centrifuge to maintain balance.

4. Cooling and Final Adjustment

- After centrifugation, allow the tube to cool for 3 minutes in a water bath at 65±2°C, but do not exceed 10 minutes.
- Adjust the contents of the tube as needed using the lock stopper and key.

5. Fat Measurement

- Read the fat percentage by observing the position of the fat meniscus at the lowest point where the fat layer separates from the acid. Ensure the butyrometer is vertical, and the reading is at eye level.
- Record the fat percentage to the nearest half of the smallest scale division.

6. Note

- For homogenized milk, centrifuge the sample again if fat separation is difficult.

Method 2: Rose-Gottlieb Method

1. **Weigh the Milk Sample**: Accurately weigh about 10 g of the milk sample and transfer it to a beaker.
2. **Add Ammonia**: Add 1.25 mL of ammonia solution (specific gravity 0.91) and shake thoroughly to mix.
3. **Transfer to Extraction Tube**: Pour the mixture into an extraction tube.
4. **Add Solvents**: Add 10 mL of ethyl alcohol, 25 mL of peroxide-free diethyl ether, and 25 mL of petroleum ether. Shake vigorously for about one minute.
5. **Separation of Layers**: Allow the mixture to separate. If emulsions form, add a small amount of alcohol to help the separation. Decant the clear ether layer into a suitable container.

6. **Repeat Extraction**: Repeat the extraction two more times using 15 mL of each solvent (ethyl alcohol, diethyl ether, and petroleum ether), and combine all extracts in the same container. Evaporate the solvents.
7. **Drying**: Dry the flask containing the extract in a hot air oven at 102±2°C for 2 hours. Cool it in a desiccator and weigh it.
8. **Re-drying and Weighing**: Heat the flask again for 30 minutes, cool it in a desiccator, and weigh it again. Repeat the heating-cooling-weighing cycle until the weight difference between two consecutive weighings is less than 1 mg.
9. **Final Fat Weighing**: Wash the extracted fat from the flask with petroleum ether, dry it, and reweigh the flask. The weight difference represents the amount of fat extracted.
10. **Blank Determination**: Perform a blank determination using the same reagents to correct for any discrepancies. The difference between two determinations by the same analyst should not exceed 0.07 g per 100 g of sample.

Calculation

Fat % (m/m) = $[(M_1 - M_2) - (M_3 - M_4)/w] \times 100$

Where,

M_1 is the mass, in g, of the fat-collecting flask and extracted matter afterdrying

M_2 is the mass, in g, of the empty fat-collecting flask

M_3 is the mass, in g, of the fat-collecting flask used in the blank test afterdrying

M_4 is the mass, in g, of the empty fat-collecting flask used in the blank test

w is the weight of the sample in g

Experiment No. – 4

AIM: To determine the milk solids in condensed or evaporated milk.

Principle

The total solids content of milk is measured by evaporating the water and comparing the weight before and after evaporation. For condensed or evaporated milk, the fat content is determined separately and subtracted from the total solids to calculate the solids-not-fat (SNF) content.

Apparatus

1. **Analytical Balance**: Accurate to 0.0001 g for precise weighing.
2. **Desiccator**: Equipped with a desiccant, such as silica gel with a humidity indicator, for cooling the dried sample.
3. **Boiling Water Bath**: Used for controlled evaporation with adjustable openings.
4. **Hot Air Oven**: Forced air type that maintains a temperature of 102 ± 2°C for drying.
5. **Flat-Bottomed Dishes**: Stainless steel, nickel, or aluminum with a diameter of 50-75 mm and height of 20-25 mm. These dishes must have removable lids.
6. **Water Bath**: For warming the milk sample, maintaining a temperature of 35-40°C.

Reagents

1. **Milk Sample**: Prepared sample, warmed to 35-40°C for uniformity.
2. **Dish and Lid**: Initially heated at 102 ± 2°C for 1 hour in the oven and then transferred to a desiccator for cooling.

Procedure

1. Preparation of the Dish

- Heat the dish with the lid in the oven at 102 ± 2°C for 1 hour.
- After heating, immediately transfer the dish to a desiccator and cool it for at least 30 minutes.
- Weigh the dish with the lid to the nearest 0.1 mg.

2. Adding Milk Sample

- Transfer 5 mL of the prepared milk sample into the cooled dish.
- Place the lid on the dish and weigh it again to the nearest 0.1 mg.

3. Evaporation of Water

- Heat the dish (without the lid) on a boiling water bath until the majority of the water has evaporated.

4. Drying the Sample

- After evaporation, wipe the underside of the dish and place it in the oven at 102 ± 2°C to dry for 2 hours. Ensure both the dish and lid are included in the oven.

- After the drying period, transfer the dish to a desiccator to cool to room temperature.

5. Weighing After Drying

- Weigh the cooled dish and lid to the nearest 0.1 mg after they have been in the desiccator.

6. Repeat Drying Process

- Place the dish and lid back into the oven at $102 \pm 2°C$ for another 1 hour of drying.
- After the second drying, place the dish in the desiccator to cool, then weigh it again.

7. Final Weighing

- Repeat the drying and weighing cycle until the mass difference between two consecutive weighings is less than 1 mg. Record the lowest mass obtained.

8. Calculation of Milk Solids

- The total solids content is determined by calculating the difference in mass between the dish before and after drying.
- The solids-not-fat (SNF) content is obtained by subtracting the fat content from the total solids weight.

Calculation

Total Solids Content (%) = $[(M_2 - M) / (M_1 - M)] \times 100$

Where,

M is the mass in g of empty dish along with lid;

M_1 is the initial mass in g of dish with lid and test portion;

M_2 is mass in g of dish with lid and dried test portion

Round the value obtained to nearest 0.01 % (m/m)

SNF Content (%) =Total Solid (%) - Fat (%)

Experiment No.- 5

AIM: To determine the protein content in the milk solids-not-fat (SNF) of condensed or evaporated milk.

Principle

This method aims to measure the true protein content in milk solids-not-fat by precipitating proteins using trichloroacetic acid (TCA). The precipitated proteins are separated from non-protein nitrogen (NPN) through filtration. Afterward, the nitrogen content in the precipitate, which represents true protein nitrogen, is measured using the Kjeldahl method.

Apparatus

1. **Conical Flask**: 25 mL capacity.
2. **Pipettes**: 10 mL and 20 mL.
3. **Filter Funnel**: 75 mm diameter.
4. **Whatman Filter Paper**: Grade 1 (nitrogen-free).
5. **Beaker**: 50 mL capacity.

Reagents

1. **Trichloroacetic Acid (TCA) Solution (15% w/v)**: Dissolve 15.0 g of TCA in water and make up the volume to 100 mL in a volumetric flask.

Procedure

A. Precipitation and Filtration

1. Weighing the Milk Sample

- Weigh approximately 3-4 g of the milk sample and transfer it to a Kjeldahl flask (500 mL or 800 mL), ensuring no sample sticks to the flask's neck.

2. Adding TCA Solution

- Add 40 ± 0.5 mL of the 15% TCA solution to the flask, swirl the mixture, and allow it to sit for 5 minutes to allow protein precipitation.

3. Filtration:

- Filter the mixture through Whatman filter paper Grade 1 into a clean conical flask. Some precipitate may remain in the Kjeldahl flask or on the filter paper.

4. Rinsing the Flask

- After filtration, rinse the flask with 10 mL of TCA solution to ensure all precipitate is transferred to the filter paper.

5. Pouring the Rinse Solution

- Mix the flask and pour the rinse solution through the same filter paper, adding the filtrate to the previously collected one.

6. Repeating the Rinsing

- Perform a second rinse with an additional 10 mL of TCA solution to ensure thorough mixing and filtration.

7. Discarding the Filtrate

- Discard the filtrate, as it is no longer needed.

8. Collecting the Precipitate

- Fold the filter paper containing the precipitate and place it back into the Kjeldahl flask, ensuring that all precipitate is retained.

B. Digestion and Nitrogen Determination

1. Adding Boiling Aids

- Add the necessary boiling aids, such as potassium sulfate (K_2SO_4), copper sulfate solution, and concentrated sulfuric acid, to the Kjeldahl flask as per the standard Macro Kjeldahl method.

2. Digestion and Distillation

- Digest and distill the contents of the flask according to the Macro Kjeldahl procedure to determine the nitrogen content of the precipitate, which represents true protein nitrogen.

C. Titration

1. Titration Procedure

- Titrate the resulting boric acid solution with 0.1 N hydrochloric acid (HCl) solution until the first appearance of a pink color, indicating the endpoint of the titration.

2. Recording the Titration Volume

- Record the burette reading to the nearest 0.05 mL for accuracy.

D. Blank Test

1. Blank Test Procedure

- Carry out a blank test using filter paper washed with TCA solution to account for any background nitrogen during the titration process.

2. Blank Value Correction

- Record the blank value from the titration to correct the final result

Calculation

The true protein nitrogen content, expressed as percentage by mass is calculated by the following formula:

$$WPN = [1.4007 \times (Vs - V_B) \times N] / W_t$$

Where,

WPN = true protein nitrogen content of the sample, expressed as a percentage by mass

V_S = volume in mL of the standard hydrochloric acid used for sample

V_B = volume in mL of the standard hydrochloric acid used for blank test

N = Normality of the standard hydrochloric acid expressed to four decimal places

W_t = mass of the test portion in g, expressed to nearest 0.1 mg.

The true protein content is obtained by multiplying the true protein nitrogen content by 6.38

Experiment No. -6

AIM: Determination of Moisture in Dried Milk Products

PRINCIPLE: The moisture content in dried milk products is determined by drying a sample at a constant temperature of 102 ± 2°C until it reaches a constant weight. The loss in weight during the drying process is used to calculate the moisture content of the sample.

Apparatus

1. **Moisture dishes with covers**: These should be flat-bottomed and made of stainless steel, nickel, or aluminum. The approximate dimensions should be 50 mm in diameter and 25 mm in depth. The lids should fit well and be easily removable.
2. **Hot air oven (forced air type)**: Capable of maintaining a constant temperature of 102 ± 2°C.
3. **Desiccator**: Contains an efficient desiccant, such as freshly dried silica gel with a hygroscopic indicator, for cooling the samples.

4. **Analytical balance**: With a readability of 0.0001 g for accurate weight measurements.

Procedure

1. Preparation of Moisture Dish

- Remove the lid from the moisture dish and place both the dish and the lid in the hot air oven at 102 ± 2°C for 1 hour to ensure they are thoroughly dried.
- After 1 hour, place the lid back on the dish, transfer it to the desiccator, and allow it to cool to 25 ± 3°C.
- Once cooled, weigh the dish accurately.

2. Sample Weighing

- Weigh approximately 1 g of the dried milk powder sample and transfer it into the moisture dish.
- Cover the dish with its lid and reweigh it to ensure accurate measurement.

3. Drying Process

- Remove the lid from the dish and place both the dish and lid in the hot air oven at 102 ± 2°C for 2 hours.
- After 2 hours, replace the lid on the dish and transfer the covered dish to the desiccator.
- Allow the dish to cool for about 30-45 minutes at 25 ± 3°C, then weigh the dish again accurately.

4. Continued Drying

- Heat the uncovered dish and lid in the hot air oven at 102 ± 2°C for an additional 1 hour to continue the drying process.

5. Cooling and weighing

- Replace the lid on the dish, allow it to cool to room temperature (25 ± 3°C) in the desiccator, and then weigh it again to check for consistency in weight.

6. Repeating the Process

- Repeat the drying, cooling, and weighing steps until two successive weighings show a difference of no more than 0.5 mg. This indicates that drying is complete and the weight is consistent.

This method ensures reliable and precise determination of moisture content in dried milk products, providing essential data for quality control and regulatory compliance.

Objective: Determination of Moisture Content in Dried Milk Products

Principle: The moisture content of dried milk products is determined by drying a known amount of the sample at a controlled temperature of 102 ± 2°C. The sample is dried until a constant weight is achieved, and the difference in weight before and after drying is used to calculate the moisture content.

Apparatus

1. **Moisture Dishes with Covers**: These should be made of materials like stainless steel, nickel, or aluminum. The diameter should be approximately 50 mm, and the depth around 25 mm. The lid must fit securely and be easy to remove.
2. **Hot Air Oven (Forced Air Type)**: Capable of maintaining a temperature of 102 ± 2°C for consistent drying.
3. **Desiccator**: Contains an efficient desiccant, such as freshly dried silica gel with a hygroscopic indicator, to cool down the sample.
4. **Analytical Balance**: With a precision of 0.0001 g for accurate weighing.

Procedure

1. Preparation of Dishes

- Place the moisture dish and lid in the hot air oven at 102 ± 2°C for 1 hour to eliminate any moisture. After this, replace the lid on the dish, transfer it to the desiccator, and allow it to cool down to 25 ± 3°C. Weigh the dish once cooled.

2. Weighing the Sample

- Accurately weigh approximately 1 g of the dried milk powder and transfer it to the prepared dish. Place the lid back on and reweigh the entire assembly to ensure accuracy.

3. Drying the Sample

- Remove the lid from the dish, placing both the dish and the lid in the hot air oven at 102 ± 2°C for 2 hours. After this period, replace the lid and return the dish to the desiccator to cool at 25 ± 3°C for 30-45 minutes. Weigh the dish again.

4. Continued Drying

- Place the dish and lid back in the oven for an additional hour at 102 ± 2°C to continue drying.

5. Final Cooling and Weighing

- After drying, replace the lid on the dish, allow it to cool to room temperature (25 ± 3°C) in the desiccator, and weigh it again to check for consistency.

6. Repeat Process

- Repeat the drying, cooling, and weighing steps until the difference between two consecutive weighings is no more than 0.5 mg, indicating that the sample has reached a constant weight.

Calculation

Moisture % by mass = $(M_1 - M_2 / M_1 - M) \times 100$

Where,

M is mass in g, of the empty dish along with lid

M_1 is initial mass in g of the dish with lid and test portion

M_2 is the mass in g of the dish with lid and dried test portion

The maximum deviation between duplicate determinations should notexceed 0.06% by mass of moisture.

Experiment No.-7

AIM: To measure the fat content in dried milk products using the Rose-Gottlieb technique.

Principle

- Ammonia solution is used to dissolve the proteins present in the sample.
- Ethyl alcohol facilitates the precipitation of proteins by reducing the water content.
- Diethyl ether and petroleum ether are employed to extract the fat from the sample.
- After the extraction, the mixed ether solvents are evaporated, and the remaining fat is weighed to determine its content.

Procedure

1. Sample Weighing

- Accurately weigh **1 g of the dried milk sample** using an analytical balance. Use a Mojonnier flask or any other suitable extraction container.

2. Adding Warm Distilled Water

- Pour 10 mL of warm distilled water (at a temperature of 65 ± 5°C) into the flask. This helps dissolve the milk powder.

3. Shaking

- Shake the flask gently to ensure that the sample dissolves completely in the warm water.

4. Adding Ammonia Solution

- Add **2 mL of ammonia solution** to the flask and shake it thoroughly.
- The ammonia dissolves the proteins, preparing them for the precipitation step.

5. Adding Ethyl Alcohol

- Introduce **10 mL of ethyl alcohol** into the flask. This causes the proteins to precipitate by reducing the water content in the mixture.
- Shake the flask well and allow the mixture to stand for a while.

6. Fat Extraction

- Add **diethyl ether and petroleum ether** in a 1:1 ratio to the flask. These solvents extract the fat from the sample.
- The fat dissolves in the ether mixture, leaving the proteins behind.

7. Separation of the Fat

- Let the mixture settle so that the fat separates from the other components, then carefully separate the fat solution from the flask.

8. Evaporating the Solvents

- Transfer the ether-fat solution to an evaporating dish or similar container.
- Evaporate the mixed ethers carefully, either under a fume hood or using a vacuum evaporator, until all solvents have been removed.

9. Weighing the Residual Fat

- After all the solvents have evaporated, weigh the remaining residue (the fat) using an analytical balance.
- The weight of the residue represents the amount of fat present in the original sample.

Calculation

Fat % (m/m) = $[(M_1 - M_2) - (M_3 - M_4) / w] \times 100$

Where,

M_1 is the mass, in g, of the fat-collecting flask and extracted matter afterdrying

M_2 is the mass, in g, of the empty fat-collecting flask

M_3 is the mass, in g, of the fat-collecting flask used in the blank test afterdrying

M_4 is the mass, in g, of the empty fat-collecting flask used in the blank test

w is the weight of the sample in g

Experiment No.-8

AIM: To determine the fat content in whole milk powder using the Gerber method.

Principle

The sample is mixed with sulfuric acid and isoamyl alcohol in a specialized Gerber tube. The sulfuric acid dissolves the protein, and the fat is released. The tube is then centrifuged, and the fat rises to the calibrated part of the tube. The amount of fat is measured as a percentage of the milk powder sample.

Apparatus

1. **Analytical Balance** (with a readability of 0.001 g).
2. **Milk Butyrometer** with a 10% scale (ISI marked).

Reagents

1. **Concentrated Sulfuric Acid** (Approximately 36.8 N).
2. **Isoamyl Alcohol** (furfural-free), with a density between 0.807 to 0.809 g/mL at 27°C.

Preparation of Gerber Sulfuric Acid

- Begin by adding a required volume of water (typically 100 mL for every 900 mL of concentrated sulfuric acid) into a Pyrex flask, which is placed in an ice-cold water bath.
- Slowly and carefully add the concentrated sulfuric acid to the water, doing so in small amounts at a time, and mix gently to avoid excessive heating.
- After cooling, measure the specific gravity of the sulfuric acid using a hydrometer. Adjust the specific gravity to the correct value (between 1.807 to 1.812 g/mL at 27°C or 1.815 to 1.820 g/mL at 20°C) by adding either more acid or water. Always maintain the same precautions to ensure safety.

- Store the prepared sulfuric acid in a glass stoppered bottle to prevent it from absorbing moisture from the air.

Procedure

1. Weighing the Sample

- Thoroughly mix the milk powder sample. Weigh out 1.69 ± 0.01 g of the milk powder using a counterpoised scoop.

2. Adding Sulfuric Acid

- Place 10 mL of prepared Gerber sulfuric acid into a butyrometer tube, ideally using an automatic dispenser. Ensure that the neck of the tube does not get wet.

3. Layering with Water

- Gently add cold water (using a wash bottle) on top of the acid to form a layer about **6 mm deep**. Allow the water to flow gently down the side of the tube.

4. Adding the Sample to the Funnel

- Insert the narrow end of a stemless funnel into the neck of the butyrometer. Carefully transfer the milk powder into the funnel. Use a camel hair brush to remove any remaining powder from the scoop.

5. Transferring Powder

- Tap the funnel lightly to ensure that most of the powder enters the butyrometer. Any remaining powder in the funnel should be transferred to the butyrometer using a glass rod and camel hair brush.

6. Adding Isoamyl Alcohol

- Add 1 mL of isoamyl alcohol to the butyrometer, ideally using an automatic dispenser.

7. Filling with Hot Water

- Add hot water (at 70°C) from the wash bottle to fill the butyrometer until it is about 5 mm below the shoulder, ensuring that all air trapped in the powder is allowed to escape.

8. Sealing the Butyrometer

- Close the neck of the butyrometer tightly with the stopper, ensuring that the contents remain undisturbed.
- If using a double-ended stopper, screw it in until the widest part is level with the top of the neck.

- If using a lock stopper, insert it until the rim touches the neck of the butyrometer.

9. Heating and Centrifuging

- Place the butyrometer in a water bath at 65 ± 2°C for at least 3 minutes, or not less than 10 minutes, ensuring that any casein particles dissolve fully.
- After heating, centrifuge the tube for 5 minutes at 1100 rpm (approximately 300 g) to allow proper separation of the fat and other components.

10. Centrifuge Alignment

- Ensure that the butyrometer tubes are arranged symmetrically within the centrifuge and spaced evenly to protect the centrifuge bearings.

11. Resting and Reheating

- After centrifugation, allow the centrifuge to come to a stop. Remove the butyrometer tubes and place them back in the water bath at 65 ± 2°C for at least 3 minutes and not less than 10 minutes.

12. Recheck if Necessary

- If a clear dividing line between the fat and the acid is not observed, or if the acid layer is unclear, repeat the temperature adjustment and centrifugation steps before proceeding.

13. Reading the Result

- Finally, take the reading from the calibrated scale on the butyrometer. This reading represents the fat content as a percentage of the whole milk powder sample.

Calculation

Note down the reading of the scale of butyrometer as follows

Scale reading = B – A

Where,

A is the reading at the bottom of the fat column

B is the reading at the top of the fat column.

Calculation the fat content in the sample as follows

Fat content (%) = Scale reading x 20/3

Experiment No. -9

AIM: To determine the titratable acidity in dried milk products.

Principle

In this method, reconstituted milk is prepared by adding water to a test portion of dried milk that corresponds to 5 g of solids-not-fat. The sample is then titrated with a 0.1 N sodium hydroxide solution using phenolphthalein as an indicator and cobalt(II) sulfate as a reference color solution. The volume of sodium hydroxide used in the titration is multiplied by a factor of 2 to express the result in terms of 10 g of solids-not-fat. The amount of sodium hydroxide required indicates the level of natural buffering substances and any developed or added acid or alkaline substances in the product.

Apparatus

1. Analytical Balance (readability 0.001 g).
2. 50 mL Burette, graduated in 0.1 mL, with an accuracy of 0.05 mL.
3. 2 mL Pipettes.
4. 50 mL Measuring Cylinders.
5. Conical Flasks (100- or 150-mL capacity), with ground necks and ground glass stoppers.

Reagents

1. Sodium Hydroxide Solution.
2. Phenolphthalein.
3. Cobalt Sulfate Heptahydrate.
4. Ethanol.

Preparation of Reagents

1. Sodium Hydroxide Standard Solution (0.1 ± 0.0002 N)

- Dissolve 0.4 g of sodium hydroxide in distilled water and make the volume up to 100 mL.
- Standardize this solution using a 0.1 N oxalic acid solution and phenolphthalein as an indicator.

2. Reference Colour Solution

- Dissolve 3g of cobalt (II) sulfate heptahydrate ($CoSO_4 \cdot 7H_2O$) in water, and dilute to 100 mL. This solution will serve as a color standard.

3. Phenolphthalein Solution

- Dissolve 2 g of phenolphthalein in 75 mL of 95% ethanol.
- Add 20 mL of water, and then titrate with sodium hydroxide solution until 1 drop causes a faint pink color to appear.
- Make the volume up to 100 mL with water.

Procedure

1. Determine Solids-Not-Fat Content

- Estimate the moisture and fat content of the sample.
- Calculate the solids-not-fat content using the formula:
- Solids-not-fat(%)=100−fat%−moisture%\text{Solids-not-fat} (\%) = 100 - \text{fat} \% - \text{moisture} \%Solids-not-fat(%)=100−fat%−moisture%

2. Weigh the Sample

- Weigh (500/a) ± 0.01 g of the test sample, where "a" is the solids-not-fat content of the sample, expressed as a percentage to two decimal places.

3. Prepare Reconstituted Milk

- Add 50 mL of water at about 20°C to the test portion and agitate it vigorously. Let the mixture stand for about 20 minutes.

4. Prepare the Reference Colour Solution

- Add 2 mL of cobalt sulfate solution to one of the conical flasks. This will create the color standard. Mix gently by swirling.

5. Using the Colour Standard

- If multiple determinations are to be made on similar products, the same color standard can be used throughout, but it should be discarded after **2 hours**.

6. Add Phenolphthalein

- To the second conical flask, add 2 mL of phenolphthalein solution and mix by gently swirling.

7. Titration

- Begin titrating the contents of the second conical flask by slowly adding 0.1 N sodium hydroxide solution from the burette while swirling the flask.
- Continue adding the sodium hydroxide until the solution turns a faint pink color that persists for about 5 seconds, similar to the color standard.

8. Complete the Titration

- The titration should be completed within 45 seconds. Record the volume of sodium hydroxide solution used, to the nearest 0.05 mL.

Calculation

The titratable acidity = 2 x V

Where,

V = volume, in millilitres, of the sodium hydroxide solution used fortitration

Experiment No.- 10

AIM: To determine the titratable acidity in dairy whitener.

Principle

The dried test portion of dairy whitener is reconstituted with water and titrated using 0.1 N sodium hydroxide solution. Phenolphthalein serves as an indicator, and rosaniline acetate solution is used as a reference color solution. The volume of sodium hydroxide solution required for the titration is dependent on the amount of natural buffering substances in the product, as well as any added or developed acidic or alkaline substances.

Apparatus

1. Analytical Balance (readability 0.001 g).
2. 50 mL Burette, graduated in 0.1 mL, with an accuracy of 0.05 mL.
3. 2 mL Pipettes.
4. 50 mL Measuring Cylinders.
5. Conical Flasks (100 mL or 150 mL capacity), with ground necks and ground glass stoppers.

Reagents

1. Sodium Hydroxide Solution.
2. Phenolphthalein.
3. Rosaniline Acetate Solution.

Preparation of Reagents

1. 0.1 N Sodium Hydroxide Solution

- Dissolve 0.4 g of sodium hydroxide in water and make up the volume to 100 mL.

- Standardize this solution using 0.1 N standard oxalic acid solution with phenolphthalein as the indicator.

2. Phenolphthalein Indicator Solution

3. Rosaniline Acetate Stock Solution

- Dissolve 0.12 g of rosaniline acetate in approximately 50 mL of rectified spirit containing 0.5 mL of glacial acetic acid.
- Make the final volume up to 100 mL with rectified spirit.

4. Rosaniline Acetate Bench Solution

- Dilute 1 mL of the stock solution to 500 mL with a mixture of rectified spirit and distilled water in equal proportions by volume.

Procedure

1. Weigh the Sample

- Accurately weigh about 1 g of the sample into two porcelain dishes.

2. Add Boiling Water

- Add 10 mL of boiling water to each dish and stir with the flat end of a glass rod until the mixture forms a smooth, uniform liquid.

3. Cool the Mixture

- Allow the mixture to cool to 25 ± 3°C. Use the contents of one dish as a blank, stirring in 2 mL of the bench solution of rosaniline acetate.

4. Add Phenolphthalein Indicator:

- Add 1 mL of phenolphthalein solution to the second dish.
- Begin titrating with the standard sodium hydroxide solution from the burette, drop by drop, while stirring continuously, until the color of the solution matches the pink tint of the blank.

5. Complete the Titration

- Stir vigorously during the titration process. The titration should be completed within 20 seconds. The process should preferably be carried out under north light or using a daylight lamp for proper color comparison.

Calculation

Titratable acidity (as % lactic acid) by mass =9 AN / M

Where,

A = volume in mL of the standard sodium hydroxide required for titration

N = normality of the standard sodium hydroxide solution, and

M = mass in g of milk powder taken for the test

Experiment No. –11

AIM: Determination of added sugar in dairy whitener.

Principle

This method involves clarifying the sample using zinc acetate and potassium ferrocyanide. Part of the sample is inverted using hydrochloric acid, and the content of reducing sugars is measured in both the inverted and non-inverted portions of the sample. The sugar content is estimated by determining the volume of an unknown sugar solution needed to completely reduce a known volume of Fehling's solution. Invert sugar reduces the copper ions in Fehling's solution, forming insoluble red cuprous oxide.

Materials And Reagents

1. Sodium Hydroxide.
2. Sucrose.
3. Methylene Blue.
4. Copper Sulphate ($CuSO_4 \cdot 5H_2O$).
5. Concentrated Sulphuric Acid (approximately 36.8 N).
6. Rochelle's Salt (Potassium Sodium Tartrate, $KNaC_4H_4O_6 \cdot 4H_2O$).
7. Zinc Acetate.
8. Potassium Ferrocyanide.
9. Concentrated Hydrochloric Acid (approximately 11.6 N).
10. Concentrated Ammonia (Specific gravity: 0.91).
11. Acetic Acid.

Preparation of Reagents

1. Sodium Hydroxide Solution

- Prepare approximately 0.1 N sodium hydroxide solution using analytical reagent-grade sodium hydroxide.

2. Stock Invert Sugar Solution

- Weigh 9.5 g of pure sucrose and dissolve it in 100 mL of water in a 1-liter volumetric flask. Add 5 mL of concentrated hydrochloric acid, let it stand for 3 days at 20-25°C, and then make up to volume with water. This solution is stable for several months.

3. Standard Invert Sugar Solution

- Neutralize 40 mL of the stock invert sugar solution with sodium hydroxide, using litmus paper, and dilute it to 100 mL.

4. Methylene Blue Indicator Solution

- Dissolve 0.2 g of methylene blue in water and dilute it to 100 mL.

5. Fehling's Solution (Soxhlet Modification)

- **Solution A**: Dissolve 34.639 g of copper sulfate in water, add 0.5 mL of concentrated sulfuric acid, and dilute to 500 mL. Filter using asbestos.
- **Solution B**: Dissolve 173 g of Rochelle salt and 50 g of sodium hydroxide in water, dilute to 500 mL, and let it stand for 2 days. Filter using asbestos.
- Combine Solutions A and B in equal volumes immediately before use.

6. Standardization of Fehling's Solution

- Mix 10 mL of each Fehling's solutions A and B in a 250 mL Erlenmeyer flask. Add standard invert sugar solution from a burette and titrate until the blue color fades. Boil for 2 minutes, add 1 mL of methylene blue, and continue titrating until the blue color disappears. Complete the titration within 3 minutes without interruption.

7. Zinc Acetate Solution

- Dissolve 21.9 g of crystalline zinc acetate in water and add 3 mL of glacial acetic acid. Make up to 100 mL.

8. Potassium Ferrocyanide Solution

- Dissolve 10.6 g of crystalline potassium ferrocyanide and make up to 100 mL.

9. Dilute Ammonia Solution

- Prepare by diluting 10 mL of concentrated ammonia solution to 100 mL with water.

10. Dilute Acetic Acid Solution

- Prepare an acetic acid solution of approximately equivalent strength to the dilute ammonia solution.

Procedure

1. Sample Preparation

- Accurately weigh 40 g of the well-mixed sample and transfer it to a 100 mL beaker. Add 50 mL of hot water (80–90°C), mix thoroughly, and transfer to a 250 mL volumetric flask, washing the beaker with successive quantities of 60°C distilled water until the volume reaches 120-150 mL.

2. Neutralization

- Allow the solution to cool to room temperature (25 ± 3°C). Add 5 mL of dilute ammonia solution, mix, and let it stand for 15 minutes.

3. Neutralize Ammonia

- Add the exact amount of dilute acetic acid to neutralize the ammonia. Mix thoroughly, and then add 12.5 mL of zinc acetate solution followed by 12.5 mL of potassium ferrocyanide solution.

4. Filtration

- Mix the solution again, make up to 250 mL with water, allow it to settle, and filter. Label this as Solution B-I.

5. Inversion Process

- Pipette 50 mL of Solution B-I into a 100 mL volumetric flask. Add 5 mL of concentrated hydrochloric acid and heat in a water bath at 65-68°C for 5 minutes, rotating the flask for the first 3 minutes.

6. Neutralization and Final Solution Preparation:

- Cool the solution, neutralize with sodium hydroxide, and mark this as Solution A-I. Make up to 100 mL.

7. Dilution

- Dilute Solution B-I to 100 mL and Solution A-I similarly. These should be diluted such that the volume required to react with 10 mL of Fehling's solution is between 15 and 50 mL.

8. Label the Solutions

- Label the diluted solutions as B-II and A-II, respectively.

Standard Method of Titration

1. First Titration

- Fill a 50 mL burette with Solution B-II. Pipette 5 mL of Fehling's Solution A and 5 mL of Fehling's Solution B into a 250 mL conical flask. Add Solution B-II from the burette to reduce all the copper ions (indicated by the fading of the blue color).
- Boil for 2 minutes, then add 1 mL of methylene blue. Continue boiling and add Solution B-II until the blue color disappears. Complete the titration within 3 minutes.

2. Repeat Titration

- Repeat the titration using **Solution A-II**, following the same steps.

Calculation

Sucrose, percent by weight $=25*W_1/W_2[2f_2/V_2-f_1/V_1]$

Where,

W_1= weight in mg of sucrose corresponding to 10 mL of Fehling's solution

W_2 = weight in g of the material taken for the determination

f_2 = dilution factor for solution A - II from A - I

V_2 = volume in mL of solution A - II corresponding to 10 mL of Fehling's solution

f_1 = dilution factor for the solution B - II from B - I

V_1 = volume in mL of solution B - II corresponding to 10 mL of Fehling's solution

Experiment No.-12

AIM: Determination of total ash in dried milk products.

Principle

The determination of total ash involves the destruction of organic matter by incinerating the sample in a muffle furnace at a high temperature (550 ± 20°C) until a constant mass of ash is obtained. The remaining ash represents the inorganic mineral content of the sample.

Apparatus

1. **Platinum or Silica Crucible**: About 70 mm in diameter and 25 to 50 mm deep.

2. **Muffle Furnace**: Capable of being controlled at 550 ± 20°C.
3. **Desiccator**: Containing an efficient desiccant (e.g., freshly dried silica gel with a hygroscopic indicator).
4. **Safety Tongs**: With long arms for handling the hot crucible.
5. **Bunsen Burner or Electric Hot Plate**: For initial heating.
6. **Analytical Balance**: Readability up to 0.0001 g.
7. **Thermal Protection Gloves**: Resistant to temperatures of 550–600°C.
8. **Hot Air Oven (Forced Air Type)**: Capable of being controlled at 102 ± 2°C.

Procedure

1. **Weigh the Sample**: Accurately weigh about 3 g of the dried milk sample into a crucible that has already been dried in a hot air oven and weighed.
2. **Initial Heating**: Gently heat the crucible on a Bunsen burner or electric hot plate at first, then strongly heat it in a muffle furnace at 550 ± 20°C until a grey ash is obtained. This process will help to burn off the organic matter.
3. **Cooling and Weighing**: After the combustion is complete, allow the crucible to cool in a desiccator. Weigh the crucible with the ash.
4. **Re-Heating**: Place the crucible back into the muffle furnace and heat it again at 550 ± 20°C for 30 minutes.
5. **Cool and Weigh Again**: After the re-heating period, cool the crucible again in the desiccator and weigh it.
6. **Repeat the Process**: Continue the process of heating for 30 minutes, cooling, and weighing until the difference between two successive weighings is less than 1 mg.
7. **Record the Lowest Mass**: Once the mass becomes constant, record the lowest mass obtained.
8. **Handling Black Particles**: If the ash still contains black particles, add 2-3 drops of preheated water (at 60°C). Break the ash and evaporate it to dryness in a hot air oven at 100–110°C. Finally, re-ash the sample at 550°C until the ash is white or slightly grey.

Calculation

Total ash % by mass = $[(M_2 – M) / (M_1 – M)] \times 100$

Where,

M_2 = mass in g, of the crucible with ash

M = mass in g, of the empty crucible

M_1 = mass in g, of the crucible with the material taken for the test and

Total ash (% on dry matter basis) = [Total ash (%)/ 100-moisture%] × 100

Experiment No. – 13

Aim: Determination of Acid Insoluble Ash in Dried Milk Products

Principle

Acid Insoluble Ash is the portion of the ash that remains insoluble in acid, which is not volatilized during incineration of the acid-insoluble residue.

Apparatus

1. Analytical Balance (Readability: 0.0001 g)
2. Platinum or Silica Crucible (Diameter: ~70 mm, Depth: 25–50 mm)
3. Muffle Furnace (Temperature control: 550 ± 20°C)
4. Desiccator (Contains efficient desiccant such as freshly dried silica gel)
5. Safety Tongs (Long arms for handling hot items)
6. Bunsen Burner or Electric Hot Plate
7. Boiling Water Bath (Temperature: 100 ± 20°C)
8. Ashless Whatman Filter Paper (Grade 42)
9. Watch-glass
10. Hot Air Oven (Temperature control: 100 ± 2°C)

Reagents

1. **Concentrated Hydrochloric Acid** (Approximately 11.6 N)
2. Silver Nitrate Solution

Preparation of Reagents

1. Dilute Hydrochloric Acid (5 N)

- Dilute 43 mL of concentrated hydrochloric acid to 100 mL with distilled water.

2. Silver Nitrate Solution (0.1 N)

- Dissolve 1.6987 g of silver nitrate in 25 mL distilled water, then make up the volume to 100 mL with distilled water.

Procedure

1. Add Hydrochloric Acid

- To the ash in the crucible (from the previous ash determination), add 25 mL of dilute hydrochloric acid (5 N).
- Cover the crucible with a watch-glass and heat it in a water bath for 10 minutes.

2. Cooling and Filtering

- After heating, allow the crucible to cool.
- Filter the contents through ashless Whatman filter paper (Grade 42).

3. Washing the Filter Paper

- Wash the filter paper with water until the washings are free of acid.
- Check by adding a few drops of the 0.1 N silver nitrate solution; continue washing until no white precipitate ($AgCl$) forms.

4. Drying the Filter Paper

- Return the washed filter paper to the crucible and place it in a hot air oven maintained at 100 ± 2°C for about 3 hours to dry.

5. Ignition in Muffle Furnace

- After drying, ignite the crucible in a muffle furnace at 550 ± 20°C for 1 hour.
- Allow the crucible to cool in a desiccator for 30-45 minutes before weighing.

6. Repetition of Ignition and Weighing

- Reheat the crucible in the muffle furnace at 550 ± 20°C for 30 minutes, cool in the desiccator, and weigh again.
- Repeat this process of heating, cooling, and weighing until the difference between two consecutive weighings is less than 1 mg.

7. Record the Lowest Mass

- Once the mass stabilizes, record the lowest mass, which represents the acid-insoluble ash.

Calculation

Acid insoluble ash, % (m/m) = [(M_2–M) /(M1 –M)]× 100

Where,

M is mass in g of the empty crucible

M_1 is mass in g of the crucible with the material taken for the test

M_2 is mass in g of the crucible with acid insoluble ash

Experiment No. - 14

AIM: Determination of WPNI of skim milk powder

Principle

The undenatured Whey Protein Nitrogen Index (WPN) is a measure of the heat treatment applied to the milk during processing to milk powder. It is the basis of the following heat classifications:

Class	WPN
High heat powder	< 1.5
Medium heat powder	> 1.5 - <6.0
Low heat powder	> 6

Apparatus

1. Spectrophotometer (UV/VIS)
2. Analytical balance (0.01 g readability)
3. Thermostatically controlled water bath (37 ± 0.5°C)
4. Electrical heat plate
5. Vibrator
6. Test tubes (25 x 150 mm, round bottom)
7. Glass funnels (short stem 65 mm, diameters 50 mm and 90 mm)
8. Pleated filter paper (S & S No. 602 for 9 cm, No. 605 for 15 cm)
9. Volumetric pipettes (1, 2, 3, 4, 5, 10, 20, 100 ml)
10. 50 ml & 500 ml Erlenmeyer flasks
11. 100 ml volumetric flask
12. 2000 ml graduated beaker
13. Stopwatch
14. Watch glasses (60 mm and 90 mm)
15. Weighing dish (disposable)

Reagents

1. HCl Solution (10 g/100 ml)

- Add 23.0 ± 0.1 ml of concentrated HCl to 77 ml of deionized water.

2. Saturated NaCl Solution

- Dissolve 1 kg of NaCl in 2 liters of deionized water. Heat below boiling point while stirring, then filter the solution through S & S No. 605 filter paper after it cools to room temperature.

Procedure

1. Sample Preparation

- Use 2.0 g of non-fat dry milk powder (adjust the sample amount if fat content is more than 1%: Amount of powder=20×100100−%fat\ text{Amount of powder} = \frac{20 \times 100}{100 - \% \text{fat}} Amount of powder=100−%fat20×100
- Fat percentage can be determined using the Gerber method).

2. Reconstituting the Milk Powder

- Dissolve 2.0 g of non-fat milk powder in 20.0 ml of deionized water in a 25 x 150 mm test tube. Use a Vibrofix to ensure the powder is completely dissolved.

3. NaCl Addition and Saturation

- Add 8.0 ± 0.1 g of NaCl to the test tube, stopper it, and place it in the water bath at 37 ± 0.5°C for 30 minutes.
- Shake the mixture 8-10 times during the first 15 minutes to ensure saturation of the milk solution with NaCl.

4. Filtration

- After 30 minutes, without cooling, shake the mixture and filter it through S & S No. 602 filter paper.
- If the initial filtrate is cloudy, re-filter it through the same filter paper.
- Collect approximately 5 ml of filtrate in a 50 ml Erlenmeyer flask.

5. Pipetting and Dilution

- Pipette 2 x 1.0 ml of the filtrate (one for sample, one for blind determination) into two 15 ml tubes.
- Add 10 ml of saturated NaCl solution to each tube. Stopper the tubes and mix by gently inverting once.

6. Acid Addition for Turbidity

- Add 2 drops of HCl solution to the sample tube (not the blind sample). Mix by gently inverting the tube twice to avoid foaming.

7. Spectrophotometer Setup

- Set the spectrophotometer to 100% transmittance using the blind sample with the wavelength set at 420 nm.
- Rinse the cuvettes with deionized water between each reading.

8. Turbidity Measurement:

- After adding HCl, wait 5-10 minutes, then invert the sample tube once more.
- Measure the turbidity by assessing the % transmittance at 420 nm using the spectrophotometer.

9. Duplicate Sample Check

- If the transmittance readings of the duplicates differ by more than 2%, repeat the process with new samples. Calculate the average of four determinations.

Results

1. Transmittance Consistency

- The transmittance readings for the duplicates should not differ by more than 2%. If they do, analyze new samples, and calculate the average of four determinations.

2. Standard Curve Application

- The % transmittance readings are then compared against a standard curve to determine the Whey Protein Nitrogen (WPN), expressed as mg WPN per gram of non-fat milk powder (NDM).

3. Moisture Content Correction:

- WPN values obtained must be corrected to a 3.16% moisture content since WPN values derived from Kjeldahl analysis are standardized to this moisture level.

Calculation

WPNI at 3.16% H_2O = [100 + (%A – 3.16) * WPN] / 100

Where,

A = % moisture in the sample

Experiment No.- 15

AIM: Determination of Insolubility Index in Dried Milk Products

Principle

Water at 24 ^{0}C(or at 50 ^{0}C if appropriate) is added to a test portion, whichis reconstituted using a special mixer. After a specified standing period, aspecified volume of the reconstituted milk or milk product is centrifugedin a graduated tube. The supernatant liquid is removed and the sedimentis re-dispersed after the addition of water at the same temperature as usedfor the reconstitution. The mixture is centrifuged and the volume ofsediment (insoluble residue) obtained is recorded.

Apparatus

1. Thermometer, capable of measuring a temperature of 24 ^{0}C or 50^{0}C with an error not exceeding ± 0.2 ^{0}C.
2. Water bath, capable of being maintained at 24.0 ± 0.2 ^{0}Cand/or 50^{0}C ± 0.2 ^{0}C, in which one or more mixing jars can be placed.
3. Mixing jar made of glass, of capacity 500 mL, as supplied for usewith the mixer.
4. Scoop, with a smooth surface, or sampling paper, black, glazed (ofdimensions 140 mm × 140 mm), for weighing the test portion.
5. Analytical balance (Readability 0.01 g).
6. Measuring cylinder made of plastic material, of capacity 100 mL± 0.5 ml.
7. Stainless steel brush, suitable for removing any residual testportion from the scoop or sampling paper.
8. Electric mixer, equivalent to that manufactured for the solubilityindex method of the American Dry Milk Institute (ADMI) with thefollowing characteristics:
   a. The 16-bladed impeller (stainless steel)shall be attached to the shaft of themixer so that the "flat" side of the impeller is underneath. The slope of the blades is upward from right to left, this is for clockwise rotation.

b. The pitch of the impeller blades shall be 30° and the horizontaldistance between the blades (around the circumference of theimpeller) shall be 8.73 mm. With usage of theimpeller, these dimensions can change and hence periodic inspectionand maintenance are essential.

c. When the mixing jar is fitted to the mixer, the length of the mixershaft shall be such that the distance from the lowest part of theimpeller to the bottom of the jar is 10 mm ± 2 mm; this means thatfor a jar of depth 132 mm the distance from the top of the jar to thelowest part of the impeller is 122 mm ± 2 mm, and to the plane of thelowest part of the impeller blades is 115 mm ± 2 mm. The impellershall also be located centrally in the jar.

d. When the mixing jar, containing 100 mL of water at 24 ^{0}C with orwithout the addition of an appropriate test portion is fitted to themixer and the mixer switched on, the impeller shall come to itsoperational fixed rotational frequency of (3600 ± 100) min^{-1} in lessthan 5 s. The direction of rotation of the impeller shall be clockwise(viewed from above). The rotational frequency of the impeller underload (as described above) shall be checked periodically with anelectronic tachometer; this is especially necessary with older modelsof the mixer, with a non-synchronous motor, whose rotationalfrequency has to be adjusted to (3600 ± 100) min−1 by means of aspeed control and a speed indicator (which may not remain accurate).

Materials and Reagents

Silicone antifoaming agent

Preparation of Reagents

Silicone Antifoaming Agent (for example an aqueous emulsioncontaining 30 percent m/m of silicone): Test the suitability of the siliconeantifoaming agent by carrying out the procedure described in the methodwithout a test portion. No more than a trace of silicone fluid (≤ 0.01 mL)should be visible at the bottom of the tube at the end of the procedure.

Procedure

1. According to whether the insolubility index is to be measured at 24 ^{0}C or at 50 ^{0}C, adjust the temperature of the mixing jar to 24.0 ^{0}C ±0.2 ^{0}C or 50.0 ^{0}C ± 0.2 ^{0}C, respectively, by keeping the jar in thewater bath, with the water level near the top of the jar, for a sufficientperiod of time. Weigh, to the nearest 0.01 g, in the scoop or on thesampling paper, a test portion of:

a) 13.00 g, in the case of dried whole milk, dried partly skimmed milk,and infant food based on either of these,

b) 10.00 g, in the case of dried skimmed milk and dried buttermilk

c) 7.00 g, in the case of dried whey.

2. Remove the mixing jar from the water bath, quickly wipe dry theoutside of the jar and, using the measuring cylinder, add to the jar100 mL ± 0.5 mL of water at 24.0 ^{0}C ± 0.2 ^{0}C or 50.0 ^{0}C ± 0.2 ^{0}C asappropriate.
3. Add 3 drops of the silicone antifoaming agent to the water in themixing jar and then transfer the test portion to the jar, using thebrush, if necessary, so that the entire test portion falls onto thesurface of the water.
4. Fit the mixing jar to the mixer, switch on the mixer, and after themixer has operated for exactly 90 s, switch it off.
5. If the mixer is of the type with a nonsynchronous motor, speedcontrol and speed indicator, bring the rotational frequency of theimpeller to 3600 ± 100 min^{-1} within the first 5 secs of the 90 secmixing period.
6. Remove the mixing jar from the mixer (allow a few secs for liquidon the impeller to drain into the jar) and leave the jar to stand atlaboratory temperature for not less than 5 min and not more than 15min.
7. Add 3 drops of the silicone antifoaming agent to the mixture in themixing jar; thoroughly mix the contents of the jar by stirring (not toovigorously) for 10 secs with a spoon spatula.
8. Immediately pour sufficient of the mixture into a centrifuge tube tofill the tube up to the 50 mL mark, that is, until the top level iscoincident with the 50 mL mark.
9. Place the centrifuge tube (counterbalanced) in the centrifuge set at20 ^{0}C to 25 ^{0}C.
10. Bring the centrifuge as quickly as possible to the rotationalfrequency, producing an acceleration of 160 g at the internal bottomof the tube, and then spin the tube at this rotational frequency for 5min.
11. Remove the centrifuge tube from the centrifuge and using a spoonspatula, remove and discard any top layer of fatty material in thetube.
12. Hold the centrifuge tube in a vertical position and remove thesupernatant liquid with a siphon fitting or a suction tube until the toplevel is coincident with the 15 mL mark, if a roller-dried product isbeing tested, or coincident with the 10 mL mark, if a spray-driedproduct is being tested, taking care not to disturb the sediment.

13. However, if it is evident that the volume of sediment exceeds 15 ml or 10 mL, respectively, discontinue the procedure at this stage andrecord the insolubility index as "greater than 15 mL" or "greater than10 mL", indicating the reconstituting temperature.
14. Otherwise, add water at 24 ^{0}C or 50 ^{0}C as appropriate to thecentrifuge tube until the top level is coincident with the 30 mL mark.
15. Completely disperse the sediment with the stirring rod, tap thebottom of the rod against the inside of the tube to collect adheringliquid and add more water at the same temperature until the top levelis coincident with the 50 mL mark.
16. Close the centrifuge tube with a rubber stopper; invert the tubequickly five times so as to mix its contents thoroughly, remove thestopper (draw the bottom of the topper across the rim of the tube tocollect adhering liquid) and spin the tube in the centrifuge for 5 minat the required rotational frequency and temperature.
17. It is recommended that the centrifuge tube be placed in thecentrifuge cup so that, when the tube is in the spinning position, thescale lines are not facing upwards or downwards but are midwaybetween these two positions. Then, if the top of the sediment issloping, the volume of sediment will be easier to estimate.
18. Remove the centrifuge tube from the centrifuge, hold the tube in avertical position against a suitable background with the top of thesediment at eye level and using the magnifying lens, read the volumeof sediment to the nearest 0.05 mL if the volume is less than 0.5 mLand to 0.1 mL if the volume is more than 0.5 ml.
19. Estimate the volume if the top of the sediment is sloping. If the topof the sediment is uneven, leave the tube in a vertical position for afew min; the top of the sediment will usually become less unevenand the volume of sediment can be read more easily. Record thetemperature of the water used for the reconstitution.

Note: Viewing the tube against either a light or a dark background, aspreferred, makes the top of the sediment more distinct.

Calculation

The insolubility index of the laboratory sample is equal to the volume, inmL, of sediment recorded. Report the result with the temperature of thewater used for reconstitution, for example as follows:

a) 0.10 mL (24 ^{0}C)

b) 4.1 mL (50 ^{0}C).

Experiment No.- 16

Aim: To Determine the Amount of Scorched Particles in Dried Milk Products

Principle

This method involves reconstituting dried milk powder with distilled water and then filtering it through a cotton disc. The resulting filtered disc, once dried, is compared visually with standard scorched particle test cards. The scorched particles present on the disc are classified based on the visual comparison with these standards.

Apparatus

1. **Standard Scorched Particles Photo Prints / Test Cards** – For comparison of scorched particles.
2. **Analytical Balance** – A torsion balance with a capacity of around 500 g and a sensitivity of 0.1 g or better.
3. **Mixer** – A Waring blender or similar device. If the blender has a worn shaft, replace it immediately to prevent contamination with metal particles, which could lead to incorrect results.
4. **Filtering Discs** – Cotton discs with a diameter of 3.8 cm (1.25 inches) or cotton pads mounted on test cards, designed for use with the aspirator-type tester.
5. **Scorched Particles Tester** – Aspirator or pressure-type tester with a filtering diameter of 3.2 cm (1.125 inches).
6. **250 mL Measuring Cylinder** – For measuring water.

Reagents

1. **Defoaming Agent** – Diglycol laurate S or Antifoam B emulsion.
2. **Water** – Sediment-free distilled water, with a temperature range of 32.5 to 41°C.

Procedure

1. Sample Preparation:

- Measure 250 mL of sediment-free water.
- Weigh the following amounts of dried milk:

- 25 g of non-fat dry milk or dry buttermilk, or
- 32.5 g of dry whole milk.

2. Mixing

- Combine the water and the weighed sample in a Waring blender jar.
- Add about 0.5 mL of diglycol laurate S to the mixture as a defoaming agent.
- Blend for 60 seconds to ensure thorough mixing.

3. Filtration

- Filter the entire mixture through a standard cotton disc using an aspirator or pressure-type tester.
- Rinse the blender jar and the tester with about 50 mL of sediment-free water, passing the rinse through the cotton disc as well.

4. Handling the Sample

- If the sample is left standing before filtration, stir the mixture vigorously before filtering. Ensure the sample is not left uncovered before the filtration step.

5. Drying the Filter Disc

- Once filtered, remove the cotton disc and place it on a scorched particle disc test card.
- Dry the disc at 30 to 40°C in a dust-free environment.

6. Comparison

- After drying, place the disc on a flat surface and view it from directly above.
- Compare the scorched particles on the disc with the standard scorched particle test cards under indirect, uniform light.
- If the disc shows more scorched particles than one standard but fewer than the next, assign it the higher letter (e.g., if the disc shows more particles than standard A but fewer than standard B, assign the result to **B**).

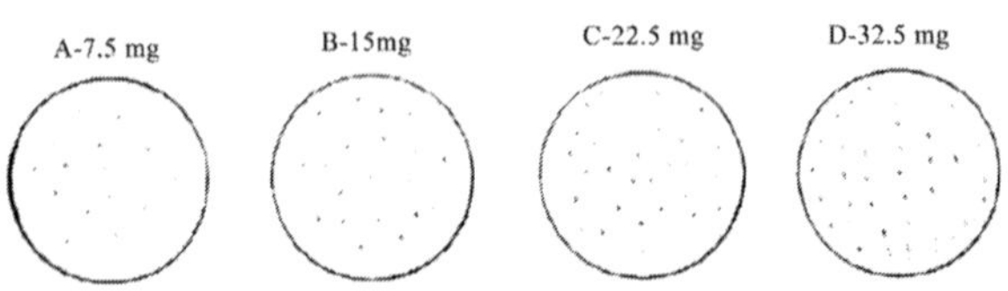

Fig. Scorched particle standards for dry milk (Source IS 13500-1992)

Experiment No.- 17

Aim: To determine the solubility percentage of milk powder in water, specifically for foods used in infant nutrition.

Principle

This method involves reconstituting a known quantity of milk powder in water under controlled conditions. The reconstituted sample is centrifuged to separate the fat from the rest of the sample. The fat layer is removed, and the remaining liquid is processed to measure the total solids in two stages: first for the whole sample, and then after the removal of the fat. The difference in total solids between the two stages indicates the solubility of the powder. The solubility percentage is then calculated based on the initial amount of milk powder used.

Apparatus

1. **Flat-bottomed Dishes** – Made from materials like aluminum, stainless steel, or porcelain, 7-8 cm in diameter and 1.5 cm in height, with tightly fitting lids.
2. **50 mL Boiling Tube** – Used to reconstitute the powder in water.
3. **Water Bath** – Maintained at 50 ± 1°C for temperature control.
4. **Hot Air Oven** – Controlled at 100 ± 2°C for drying.
5. **Centrifuge** – Capable of operating at 770 × g.
6. **Analytical Balance** – With a sensitivity of 0.01 g for accurate weighing.
7. **Aluminum Dish with Lid** – For determining total solids in the sample.
8. **Refrigerator** – To cool the sample at 4-7°C.

Procedure

A. Reconstitution of Milk Powder

1. Accurately weigh 4 g of the milk powder into a 50 mL boiling tube.
2. Add 32 mL of water at 50 ± 1°C into the tube. Cork the tube and shake for 10 seconds.
3. Place the tube in a water bath at 50 ± 1°C for 5 minutes.
4. Shake the tube for 1 minute to ensure proper mixing.

B. Removal of Fat

1. Transfer the reconstituted milk into a 25 mL centrifuge tube.
2. Centrifuge the sample for 10 minutes at 2000 rpm with a radius of 17 cm (giving a force of 770 g).

3. Cool the centrifuge tube in a refrigerator or on ice until the fat solidifies (avoid freezing the milk).
4. Carefully remove the solidified fat layer using a spatula.
5. Allow the remaining milk to reach room temperature (27 ± 1°C).
6. Break up the milk deposit with a glass rod, cork the tube, and shake it vigorously until the mixture becomes uniform.

C. Determination of Total Solids

1. Take 2 mL of the homogeneous liquid and transfer it to a pre-weighed dry tared aluminum dish (Dish No. 1).
2. Weigh the dish with its lid and set it aside.
3. Centrifuge the tube again for 10 minutes at 2000 rpm with a radius of 17 cm (770 g).
4. Pipette 2 mL of the upper supernatant (without disturbing the sediment) into a second aluminum dish (Dish No. 2).
5. Weigh the second dish with the lid.
6. Remove the lids and place both dishes side by side on a water bath until the liquid evaporates.
7. Transfer the dishes to an air oven at 100 ± 2°C for 90 minutes.
8. After drying, cover the dishes and place them in a desiccator to cool before weighing them individually.

Calculation

Solubility, percent by weight =[M4 × M1 ×100] / [M3 × M2]

Where,

M_1 is Mass in g of the liquid taken immediately after the removal of fat

in dish No.1

M_2 is Mass in g of the supernatant liquid taken in dish No. 2

M_3 is Mass in g of total solids in dish No. 1

M_4 is Mass in g of total solids in dish No. 2

Experiment No. -18

AIM: Determination of Milk Protein in Colostrum

Principle

The method is used for analysis of true protein content by estimating thenitrogen content obtained only from proteins and not from non-proteinnitrogen (NPN). In this method, the protein in milk sample is precipitatedby the addition of TCA such that the final concentration of TCA in themixture is approximately 12%. The precipitated proteins are filtered andthe filtrate will contain the NPN contents. The nitrogen content of theprecipitated protein is then determined by Kjeldahl method whichrepresents true protein nitrogen content of milk.

Apparatus

1. Conical flask: 25 mL capacity.
2. Pipettes: 10, 20 mL capacities.
3. Filter funnel: 75 mm diameter.
4. Whatman filter paper: Grade 1 (nitrogen free).
5. Beaker: 50 mL capacity

Reagents

1. Trichloroacetic acid (TCA)

Trichloroaceticacid(TCA)solution(15%,w/v):Dissolve15.0goftrichloroacetic acid (CCl3COOH) in water and dilute to 100 mL in avolumetric flask.

Procedure

Weigh quickly about 3-4 g of the prepared colostrum sample and transferto a 500 or 800 mL Kjeldahl flask/digestion tube taking care to see that noportion of the sample clings to the neck of the flask/tube.

A. Precipitation and Filtration

1. Add 40 ± 0.5 mL of TCA solution to the flask and swirl to mix thecontents. Let the flask stand for approximately 5 min to allow theprecipitate to settle.
2. Pour the mixture from the Kjeldahl flask through a Whatman filterpaper Grade1 and collect the entire filtrate in a clean, dry conical flask.Some of the precipitate will remain in the Kjeldahl flask and some willbe collected on the filter paper. It is not necessary to remove all of theprecipitate from the flask.

3. Immediately after pouring the mixture (and so as not to allow anyprecipitate to dry on the neck of Kjeldahl flask), add, by means of apipette, 10 mL of TCA solution.
4. Use the acid solution to also rinse any precipitate from the neck of theflask down on to the bottom, swirl to mix the contents.
5. Pour the mixture from the flask through the same filter paper, addingthe filtrate to that collected previously.
6. Immediately rinse the neck of the flask with a further 10 mL of TCAsolution again.
7. Swirl to mix the contents and again pour the mixture from the flaskthrough the same filter paper, adding the filtrate to that collectedpreviously. This filtrate shall be clear and free of particulate matter.

   Note: At this point, the filtrate is no longer needed and may be discardedin an appropriate manner.
8. Wearing gloves, carefully remove the filter paper from the filter funneland fold the filter paper to enclose the precipitate.
9. If any precipitate remains on either the inner or outer lip of the Kjeldahlflask, wipe with the folded filter paper so that any precipitate adheresto the paper and then drop the filter paper into the Kjeldahl flask.

B. Digestion and Distillation

1. Add the appropriate amount of boiling aids, K_2SO_4, copper (II) sulfatesolution and concentrated sulphuric acid as specified in the MacroKjeldahl method for the determination of total nitrogen content in milk.
2. Determine the nitrogen content of the precipitate and filter paper by theMacro Kjeldahl method.

C. Titration

1. Titrate the boric acid receiving solution with standard hydrochloric acidsolution (0.1 N) to the first trace of pink colour.
2. Take the burette reading to at least the nearest 0.05 ml. A lighted stirplate may aid visualization of the end point.

D. Blank Test

Carry out a blank test as described above. Replace the test portion by afilter paper washed with TCA solution and record the blank value.

Calculation

The true protein nitrogen content, expressed as percentage by mass iscalculated by the following formula:

$WPN = [1.4007 \times (Vs - V_B) \times N] / W_t$

Where,

WPN = true protein nitrogen content of the sample, expressed as apercentage by mass

V_S = volume in mL of the standard hydrochloric acid used for sample

V_B = volume in mL of the standard hydrochloric acid used for blank test

N = Normality of the standard hydrochloric acid expressed to four decimalplaces

W_t = mass of the test portion in g, expressed to nearest 0.1 mg.

The true protein content is obtained by multiplying the true proteinnitrogen content by 6.38

Experiment No.-19

Aim: To determine the presence of melamine in milk and infant formula using liquid chromatography-tandem mass spectrometry (LC-MS/MS).

Principle

The method for detecting melamine and its residues in milk and infant formula involves a series of extraction, filtration, and centrifugation steps. The sample is initially extracted with a 2.5% aqueous formic acid solution. The process includes filtration and dilution, followed by centrifugation to separate different components of the milk matrix. Both melamine and cyanuric acid are detected using Hydrophilic Interaction Liquid Chromatography (HILIC) and tandem mass spectrometry. The concentration of melamine is determined by creating a calibration curve using fortified control matrices, which undergo the same extraction process as the test samples.

Apparatus

1. **Liquid Chromatograph**: Binary LC pump for precise mixing at low flow rates and quick response to mobile phase gradient changes.
2. **Column**: ZIC-HILIC (2.1 × 150 mm, 5 µm, 200 Å) or equivalent.
3. **Mass Spectrometer**: Triple quadrupole mass spectrometer capable of meeting system suitability requirements.

4. **Analytical Balance**: With a readability of 0.0001 g for precise measurement.
5. **Centrifuge**: Capable of 4000 rpm (approximately 3750 g) with 50 mL tubes.
6. **Microcentrifuge**: Capable of 13,000 rpm (16,100 g) with 1.5/2 mL tubes.
7. **Mixers and Shakers**: For thorough sample mixing, both single and multi-tube vortex mixers are used.
8. **Ultrasonic Bath**: Equipped with a timer and heater for efficient sample sonication.
9. **Centrifuge Tubes**: Disposable polypropylene 50 mL tubes with caps and 1.5 mL microcentrifuge tubes.
10. **Syringe Filters**: Polyvinylidene fluoride (PVDF) filters, 13 mm, 0.22 μm for sample filtration.
11. **Syringes**: Polypropylene 3 mL syringes for transferring solutions.

Reagents

1. **Melamine**: Standard for comparison and calibration.
2. **Acetonitrile (LC grade)**: High-purity solvent for the chromatography.
3. **Formic Acid (Reagent grade >95%)**: Used for acidifying the extraction solvent.
4. **Water (LC grade)**: Purified to > 18 MΩ resistivity or equivalent.
5. **Ammonium Formate (Purity > 97%)**: Used in mobile phase preparation.

Preparation of Reagents

1. **0.1% Formic Acid in Water**: Add 1 mL of formic acid to 1 L of LC-grade water and dilute.
2. **Mobile Phase A**: Mix 50 mL of 0.1% formic acid in water with 950 mL of acetonitrile to make 1 L.
3. **Mobile Phase B**: Mix 500 mL of 20 mM ammonium formate and 500 mL of acetonitrile to make 1 L.
4. **2.5% Formic Acid in Water**: Add 25 mL of formic acid to 1 L of LC-grade water.
5. **20 mM Ammonium Formate**: Dissolve 0.63 g of ammonium formate in 500 mL of LC-grade water.

Procedure

1. Sample Preparation

1. Weigh 2.0 ± 0.1 g of milk or infant formula into a 50 mL polypropylene centrifuge tube.
2. Add 14 mL of 2.5% formic acid solution to the sample. Seal the tube tightly.
3. Shake the tube for 15-30 seconds to dissolve the sample. Use a vortex mixer if necessary.
4. Sonicate the mixture in an ultrasonic bath, then mix on a multi-tube vortex mixer for 30 minutes.

2. Centrifugation

1. Centrifuge the sample at 4000 rpm (3750 g) for 10 minutes at 25 ± 3°C.
2. Transfer approximately 1.4 mL of the supernatant into a 1.5 mL microcentrifuge tube.

3. Second Centrifugation

1. Centrifuge the microcentrifuge tube at 13,200 rpm (16,100 g) for 30 minutes.
2. Transfer the supernatant into a 3 mL syringe and filter it through a 13 mm, 0.22 μm PVDF filter into another microcentrifuge tube. Some formulations may require additional force or a second filtration step to achieve a clear solution.

4. Storage and Final Preparation

1. The filtered aqueous extract can be stored at 5-10°C for future dilutions.
2. Vortex mix the sample for 30 seconds and centrifuge it again at 13,200 rpm (16,100 g) for 30 minutes.
3. Transfer the supernatant to a 2 mL autosampler vial, ensuring that no precipitate is included.

Calculation

Use external standard calibration. The calibration curve should not includethe origin, but does include a matrix blank with a concentration of 0. Exportthe processed data into Microsoft Excel or equivalent spreadsheet programfor further calculations:

Recovery (%) = calculated from extracted calibration curve

Matrix effect(%)= 100 x Post-fortified sample/solvent standard (sameconcentration)

The limit of quantification (LOQ) for each analyte is defined as theconcentration of the lowest calibration standard used, or the lowest calibrationstandard which shows > 10-fold higher response than background signals innegative control sample.

Experiment No.-20

Aim: To determine the crude fiber content in both milk-cereal-based complementary food and processed cereal-based complementary foods.

Principle

The process involves the acid and alkali treatment of the sample. After grinding and defatting (if necessary), the sample is first treated with a sulfuric acid solution, which dissolves the soluble components. The insoluble residue left after acid treatment is then washed and treated with a sodium hydroxide solution. After washing and drying, the residue is weighed. The final step is incineration, where the mass loss is measured to determine the crude fiber content of the sample.

Apparatus

1. **Grinding Device**: Suitable for the sample type, ensuring no excessive heating or significant moisture loss during grinding.
2. **Burette**: Graduated to 0.1 mL with an accuracy of 0.05 mL for precise liquid measurement.
3. **Sieve**: A metal wire sieve with 1 mm aperture, complying with ISO 3310/1 standards.
4. **Hot Air Oven**: Forced-air type oven with an adjustable temperature setting of **130 ± 2°C**.
5. **Wide-Mouthed Vessel**: Minimum capacity of 600 mL, equipped with a reflux condenser or similar setup.
6. **Heating Device**: A hotplate with a magnetic stirrer capable of maintaining a gentle boil for 200 mL of reagents.
7. **Incineration Dish or Filter Crucible**: A suitable container (25–50 mL) for residue incineration.
8. **Muffle Furnace**: Equipped with temperature control and air circulation, set to 550 ± 25°C.

9. **Desiccator**: Contains a drying agent (e.g., silica gel) for cooling the sample.
10. **Separating Device**: For isolating the insoluble residue after boiling.
11. **Analytical Balance**: With a readability of **0.0001 g** for accurate weighing.

Reagents

1. **Concentrated Sulfuric Acid**: Approximately 36.8 N for the acid treatment.
2. **Sodium Hydroxide**: For the alkali treatment.
3. **Acetone or 95% Ethanol**: Used for washing the residue after treatment.
4. **Concentrated Hydrochloric Acid**: Approximately 11.6 N, if needed for neutralizing carbonates.
5. **Extraction Solvent (n-Hexane, Petroleum Ether, Diethyl Ether)**: Used for extracting fats if present in the sample.
6. **Filter Aid**: For assisting with the separation of residues.
7. **Anti-foam Agent**: (e.g., silicone emulsion) to reduce foaming during boiling.
8. **Anti-bumping Agent**: Glass beads, used to prevent violent boiling.

Preparation of Reagents

1. **Sulfuric Acid Solution (0.255 N)**: Prepare by diluting 12.68 mL of concentrated sulfuric acid to 1000 mL with distilled water.
2. **Sodium Hydroxide Solution (0.313 M)**: Dissolve 12.5 g of sodium hydroxide in 1000 mL of distilled water.
3. **Acetone or Ethanol**: Use for washing the residues.
4. **Hydrochloric Acid Solution (0.5 mol/L)**: For products containing carbonates, prepare by adding 21.6 mL of concentrated hydrochloric acid to 500 mL of distilled water.
5. **Extraction Solvent**: Choose a solvent such as n-hexane or diethyl ether for fat extraction.

Procedure

A. Grinding the Sample

1. If the sample does not pass through the sieve, grind it until it can pass through the 1 mm aperture. If the sample is moist, determine its dry matter content before grinding.

B. Extraction of Fatty Substances

1. For samples with fat content above 1%, extraction is required.
2. Use appropriate solvents such as n-hexane or diethyl ether for fat removal before proceeding with acid treatment.

C. Acid Treatment

1. Weigh about 3g of the sample, which should contain over 1% crude fiber.
2. If necessary, dry the sample before beginning the procedure.
3. Add the sample to a vessel and add filter aid, anti-foam, and anti-bumping agents.
4. Add 200 mL of sulfuric acid solution and heat to 95–100°C. Boil gently for 30 ± 1 minutes.
5. Periodically swirl the vessel to ensure uniform heating and prevent particles from sticking to the walls.
6. After boiling, add 50 mL of cold water and separate the insoluble residue using a separating device.
7. Wash the residue with hot water (95–100°C) until the filtrate is neutral.

D. Alkaline Treatment

1. Return the washed residue to the vessel, adding anti-foam and anti-bumping agents.
2. Add 200 mL of sodium hydroxide solution and heat to 95–100°C. Boil gently for 30 ± 1 minutes.
3. After boiling, add 50 mL of cold water and separate the insoluble residue.
4. Wash with 25 mL of sulfuric acid solution followed by hot water to remove any remaining soluble substances.

E. Drying

1. Dry the residue in the oven at 130 ± 2°C until constant weight is achieved.
2. After drying, allow the residue to cool in a desiccator and weigh it to the nearest 0.5 mg.
3. Repeat the drying process until two consecutive weighings differ by less than 1 mg.

F. Incineration

1. Incinerate the dried residue in the muffle furnace at 550 ± 25°C until it reaches a constant mass.
2. After incineration, cool the residue in a desiccator and weigh it again to the nearest 0.5 mg.

Number of Determinations

- At least **two determinations** should be carried out on the same sample to ensure accurate results.

Blank Test

- If asbestos is used as a filter aid, perform a blank test under the same conditions to account for any potential contamination.

Calculation

Crude fibre content relative to product as received:

The crude fibre content, expressed as percentage by mass relative to theproduct as received, is given by the following formula:

a) for products not requiring grinding:

   Crude fibre = $[M_1 - (M_2 + M_3)]$ x100/Mo

b) for products requiring grinding:

   Crude fibre = [M1 – (M2 + M3)] × (100/Mo) × (100/M"s) × Ms/100

Where,

M_0 is mass, in g, of the test portion

M_1 is total mass, in g, of the dry residue and its support after drying

M_2 is total mass, in g, of the dry residue and its support after incineration

M_3 is difference in mass, in g, observed during the incineration process inthe blank test, taking into account the quantity of filter aid used;

Ms is dry matter content, expressed as a percentage by mass, of the productas received,

M"s is dry matter content, expressed as a percentage by mass, of the testsample.

Crude fibre content relative to the dry matter content of the product:

The crude fibre content, expressed as a percentage by mass relative to thedry matter content of the product, is given by the following formulae:

Crude fiber = [M1 – (M2 + M3)] × (100/Mo) × (100/M"s)

Where,

M_0 is the mass, in g, of the test portion

M_1 is the total mass, in g, of the dry residue and its support after drying

M_2 is the total mass, in g, of the dry residue and its support afterincineration

M_3: is the difference in mass, in grams, observed during the incinerationprocess in the blank test, taking into account the quantity of filter aid used

M"s is the dry matter content, expressed as a percentage by mass of thetest sample.

Experiment No.-21

Aim: To determine the total carbohydrate content in both milk-cereal-based complementary food and processed cereal-based complementary food.

Principle

The total carbohydrate content is calculated by subtracting the sum of moisture, fat, protein, and ash content from 100. This approach assumes that the remaining components, once moisture, fat, protein, and ash are accounted for, consist entirely of carbohydrates.

Apparatus

Since this method relies on calculation, it requires the use of standard apparatus for determining:

- Moisture content
- Fat content
- Protein content
- Ash content

Reagents

This method does not require specific reagents for carbohydrate determination. However, the reagents used to determine moisture, fat, protein, and ash contents in the sample are necessary.

Procedure

1. **Moisture Determination**: Measure the moisture content of the sample using an appropriate method such as oven drying or Karl Fischer titration.

2. **Fat Determination**: Use methods like Soxhlet extraction to determine the fat content of the sample.
3. **Protein Determination**: Determine the protein content using a method like Kjeldahl or Dumas combustion.
4. **Ash Determination**: Incinerate the sample in a muffle furnace at a high temperature to determine the ash content.

Calculation

Total carbohydrate including sucrose, dextroseand dextrin, maltose or lactose percent by weight = 100 – (A+B+C+D)

Where

A = Percent by mass of moisture

B = Percent by mass of total protein

C = Percent by mass of fat

D = Percent by mass of total ash

References

De, S.(1982). Outlines of Dairy Technology. Oxford University Press, New Delhi.

Fox, P.F., Uniacke-Lowe, T., McSweeney P.L.H. and O'Mahony, J.A. (2015) Dairy Chemistry and Biochemistry. (Second Edition). Springer. ISBN:9783319148915

FSSAI Manual of Methods of Analysis of Foods Dairy and Dairy Products,(2022).

Mann, B., Arora. S., Sharma, R., Seth, R., Sharma, V., Bajaj, R., Gandhi, K., Rao, P. S. and Singh, R. A Laboratory Manual on Analysis of Milk Products, Dairy Chemistry Division ICAR

Mathur, M.P., Roy, D.D. and Dinakar, P. (2019) Textbook of Dairy Chemistry. Indian Council of Agricultural Research, New Delhi. ISBN: 9788171640461

Varnam, A.H. and Sutherland, J.P.(1994). Milk and Milk products (Technology, Chemistry and Microbiology), Springer

Index